Educating Farmers about Agriculture

Raghunath Singh
Principal Scientist (Retd.)
Central Arid Zone Research Institute (CAZRI)
Jodhpur (Raj.)

SCIENTIFIC
PUBLISHERS

Published by
SCIENTIFIC PUBLISHERS (INDIA)
5 A, New Pali Road, P.O. Box 91
Jodhpur 342 001, INDIA
E-mail: info@scientificpub.com
Website: http://www.scientificpub.com

ISBN: 978-93-88449-24-3
eISBN: 978-93-88449-28-1

Printed in India

Foreword

Dr. R.N. Singh has made a praiseworthy attempt to put together the results of a wide variety of attempts by different technology generation and dissemination agencies of the Government of India. He has summarized the results and attempted to draw broader inferences thereof. All academicians, scholars and extension policy makers will find this summation of the results of notable studies evaluating the outcomes of the technology transfer strategies.

I personally have known Dr Singh for the last nearly half a century. I have always been amazed by his dedication and energy in taking up challenging projects and putting his best efforts in achieving his goals. There is a fine blend of academic interest without ever losing sight of practical relevance, especially in programs that ultimately promise long-term interest to the farming community.

I sincerely hope that this publication will be of value to all whom it is addressed.

Dr. T.K. Chakravarty

Former Director,
Applied Behavioural Sciences
National Institute of Agricultural Extension Management,
Hyderabad

Preface

Education brings desirable change in knowledge, attitude and skill of human beings. It is the most potent force for molding a society into its desired form. Hence, farmers in India also needs to be trained and educated with advanced and modern methods of farming. In the post-independence period institutions were established for farmers' education in agriculture technology. Planning and implementation of the programmes were entrusted to government organisations, research institutes and voluntary agencies.

The present book reports an overview of the studies of four such programmes such as (National Demonstration, Lab to Land, Operational Research Project and *Krishi Vigyan Kendra*) undertaken by Indian Council of Agricultural Research, New Delhi. In these pioneering efforts agriculture scientists' educated farmers about scientific know-how of agriculture developed at government research institutes and the agricultural universities. All programmes were based on conceptual frameworks with the aim of replicating such modern methodology across the country.

The book deals with genesis of the selected programmes, the extension methods used impact of the programmes and suggestions for improvement. The programmes reviewed gave very positive results. A wide gap existed in the yield of crops between the yield obtained in demonstration plots and the farmers' fields. The farmers covered by the programmes had better knowledge and adoption of improved agricultural practices than the villages that remained outside their coverage. The programme proved to be quite profitable to the participating farmers.

However, there exists great scope for improving agricultural productivity. This happens because of two basic factors. First and foremost, adoption of innovations in agricultural technology among farmers is still quite low compared to the extent of possibility. Secondly, technology is not static; it is always in the process of refinement and up-gradation as the stream of fresh knowledge flows in. It is, therefore, imperative that all-out efforts be made for effective farmers' education. Adequate knowledge of modern farming techniques will motivate and inspire farmers to try and practice whatever is feasible and required in their situations. Of course, the development agency has to be ready with effective agricultural input supply system, remunerative price for farm produce and marketing facility their produce. A vigorous effort by development personnel is needed to facilitate adoption of techniques by farmers because it is the farmer who cultivates the field and takes decision for adoption of agricultural technology.

A foundation course at school level for imparting rudimentary knowledge of agriculture will go a long way towards acceptance of scientific technology by farmers later on.

The author expresses his deep sense of gratitude to Dr. M. Joshi, Associate Professor, Jaynaraian Vyas University, Jodhpur for encouragement while writing the book.

The author expresses gratitude to the librarians of Maharana Pratap University of Agriculture, Udaipur, Mahatma Phule Krishi Vidyapeeth, Rahuri and Banaras Hindu University, Varanasi for allowing me to use their libraries.

Dr. R.N. Singh

Contents

Chapter - 1

Introduction to Farmer's Learning

THE FARMER

Agricultural advancement is vital for development of the country. The policies of the government, work of scientists, efforts by extension workers, credit institutions, input agencies, marketing agencies and physical labour done by farmers, all are involved with distinct roles in the process of agriculture development. Farmers naturally occupy the pivotal position in this scheme. There are several important things which a farmer does on his fields which includes ploughing the field, preparation of seed beds, applying of manure to fields, sowing of crop, application of fertilizer, weeding, irrigating the crops, control measures for pest and diseases, harvesting, threshing, winnowing and storage of food grain. The farmer has to tend his fields in extreme weather conditions. With respect to animal husbandry, his work includes collection of fodder, chaffing, preparation of concentrate feed for cattle, feeding and watering them, cleaning and washing the cattle shed, cleaning and bathing animals, milking, animals grazing, collection of cow dung for manure for the farm and preparation of cow dung cakes.

Farmers are also managers of the farm as they take decision about various agriculture activities to be undertaken at the farm. These include choosing the right crops that are to be sowed in each field, choice of livestock distribution of available labour for different tasks. They also take care about seed improvement, fertilizers, insecticide or new implements to be used. When the crop is finally harvested, they decide how much product is to be sold, when to sell it and whom to sell at what price.

The farmers do agricultural operations with their family members including women and men. Farm women contribute much in agriculture. In the small and marginal farmer's families, women are involved in many farm operations either in their own fields or at others as hired labour. The rate of participation of women has been 58 per cent in marginal farms, 53 per cent in small farm and 54 per cent in medium farms (Saikia, 1985). In Rajasthan women's joint participation in agriculture operations was up to 75 per cent (Tewari *et al.*, 1996). Nataraju and Lovely (1989) indicated that farm women's participation was crucial in various operations such as sowing/transplanting (86%), weeding (84%), storage of grain (78%), land preparation (72%), cleaning seed for sowing (70%), harvesting (64%) and threshing and winnowing (62%).

In spite of the hard work the farmers do not get remunerative price of their agriculture produce. Price of agriculture produce is low at the time of harvest of crops when they need to sell it. They cannot keep big stocks as they need money to meet family expenses. The shop or businessmen with deep pockets take advantage of this and purchase the produce at low price, stock it and raise its price later. Farmers of rainfed area are vulnerable and run high risk in agriculture because rainfall is non-dependable in both its timing and volume. Very often they are not able to sow the seed in time. The consequences are: the crops do not grow well, there is no flowering, grains are of small size and thus crops give low yield. Erratic rainfall also affects application of chemical fertilizer to the crop because chemical fertilizer is applied in conditions of low moisture. They face difficulty in spraying pesticide due to lack of water. Sometimes there are frequent droughts in the area. Rain can fail at sowing time or during the vegetative period or at flowering or grain setting and therefore, it is always risk to apply high cost inputs – chemical fertilizer and pesticide. There are instances when the crop is not sown because drought occurs at sowing time and thus it causes a collapse of agriculture. In severe failure of rain the crop is badly affected. The growth of the crop is stunted, the leaves become pale, there is no grain setting and only fodder is available. In moderate drought conditions after sowing, there is low crop yield and little fodder. The farmers do not even recover the cost incurred in cultivation got from the entire effort of sowing and cultivation. At such times the government has to come to their rescue and start relief work to minimize the effect of drought. This makes a big drain on the financial resource of the government.

PROGRAMME FOR FARMERS' EDUCATION FOR SCIENTIFIC KNOW- HOW OF AGRICULTURE

Knowledge of scientific know-how of agriculture developed by research institutes among the farmers is indeed a necessary precondition for adoption of new method for agricultural advancement. There can be no adoption of scientific know-how on depth without its knowledge. It provides a cognitive foundation for further process of evaluation, trial and actual adoption. The knowledge of scientific know-how of agriculture is a variable that can be significantly manipulated among farmers. Education brings desirable change in knowledge, attitude and skill of farmers regarding scientific know-how of agriculture.

Soon after independence the Government of India launched ***Community Development Programme*** on 2 October, 1952 on the anniversary of Mahatma Gandhi's birthday. The programme has been hailed as the world's largest experiment for human development. The objective of the programme, which was incorporated in the First Five Year Plan, was to quickly increase agriculture production, to accelerate the development of villages and small scale industries; so as to secure the fullest development of human and natural resources of the area. A new administration unit ***'The Community Development Block'*** was created as a primary unit of development. Each CD lock was constituted of approximately 100 villages with population coverage of sixty to eighty thousand people and is manned by a team of Block Development Officer (BDO), assisted by subject matter specialists and village level workers (VLWs). The job of the personnel of community development was completely different from the executive functions of the officials of the colonial period. They worked as educators to the villagers telling them about the government programmes that covered several aspects of village life such as, agriculture, animal husbandry, health, sanitation, cooperative and communication etc. Extension methods were used to impart knowledge and skills and stimulate interest in the government programmes.

The studies that were undertaken during initial period of community development revealed that there was not much people's participation in the programme. In view of it, the ***Committee on Plan Projects***, a high powered body under the chairmanship of the Union Minister of Home Affair, appointed in 1957, a ***Study Team*** to examine the success and efficiency of the programme. The Study Team was headed by the Balwantray Mehta in 1957. The recommendation of the Study Team was accepted by National Development Council. The Panchayat was given a legal status and entrusted with development functions for evoking villager's participation in the programme.

The Government of India was concerned about the slow growth of agriculture. The steep fall in crop production in the drought year of 1958-59 forced attention on the seriousness of the food situation. Imports of food grain had to be stepped up substantially. In these circumstances, the Government of India invited a team of agriculture experts from U.S.A to make a careful study of the Indian agriculture and make recommendations for future action. The team visited India early in 1959 and submitted its report entitled *"India's Food Crisis and Steps to Meet it"*. The team recommended a number of measures for reaching the desired goals of food production. One of the important recommendation was that instead of spreading development efforts more or less on a uniform basis throughout the country without getting any striking results, intensive efforts for agriculture production should be undertaken with concentration of all technical advanced practices, manpower and resources in selected areas which had optimum condition for stepping up production without at the same time affecting the national efforts in other areas. The Government of India accepted in general the recommendations made by the team and ***Intensive Agricultural District Programme (IADP)*** popularly known as Package Programme was begun in selected districts during 1960-61. The programme covered 28 districts of India. It envisaged the selection of favorable areas which had the maximum irrigation facility and minimum of natural hazards. These areas were to be given all ingredients of agriculture production such as adequate supply of fertilizer, credit and water. The aim was to bring an integrated approach to agricultural production in the areas which were more responsive to such efforts. The staff was also strengthened. The success of the IADP in its initial years encouraged its expansion in other promising districts. During 1964, the ***Intensive Agricultural Area Programme*** (IAAP), was started in 114 districts and extended later to other 150 districts. The IAAP also followed the package approach, but the staffing pattern in these districts was less intensive than IADP districts.

Both the IADP and IAAP were concerned with promotion of intensive agriculture. They operated within limitation set by then the existing crop varieties which had low fertilizer response. A few Mexican dwarf wheat varieties were tried out on selected basis in 1964-65 and also an exotic variety of paddy, Taichung Native 1, was tried and the results were quite encouraging. The draft outline of the Fourth Plan therefore stressed the need for evolving a new approach for boosting up agricultural production over a short span of time and in 1965 a new programme was started by the Government of India for increasing cultivation of high yielding varieties. This was known as the ***High Yielding Variety Programme*** (HYV). Another centrally sponsored scheme of Farmers' Training and Education was launched in the year 1966-67 with a view to primarily supporting the High Yielding Variety Programme. Emphasis was placed on rice, wheat, maize, sorghum and pearl millet under the intensive agriculture. During late sixties the package approach was extended to certain commercial crops: tobacco, sugarcane, sugar beet, cotton, jute, oilseeds and fruit crops.

The World Bank emphasized strengthening of extension system for effectively educating farmers regarding agriculture technology. The World Bank also helped in the introduction of **Training and Visit (T&V)** system of agricultural extension in India. It was first started in Chambal Command Area of Rajasthan, Madhya Pradesh and Poochapad Command Area of Andhra Pradesh in 1974-75. It was adopted by many states of India with the assistance of the World Bank. In the T&V system the VLWs devoted all their time to educate the farmers about agriculture technology. They were not given regulatory functions, input supply, credit or other activities relating to the rural development programmes. The jurisdiction of the VLWs was 500 farm families in irrigated area and 800 farm families in dryland areas. The area was divided into eight groups, each of which was visited by the VLW on a fixed day each fortnight. Ten per cent of the farmers in each group were selected by the VLW as contact farmers. He used to meet the contact farmers on scheduled days and advice and guide them on agriculture technology which needed to be adopted at each stage of cultivation, even advising them what they needed to do the next fortnight. There was a regular training of the VLW. In the training sessions the VLWs were intensively trained in three or four agriculture technologies.

Several Research Institutes are working under Indian Council of Agricultural Research (ICAR), New Delhi – a central body of unique importance. The ICAR, New Delhi established *Krishi Vigyan Kendra (KVK)* for education of farmers in the field of agricultural technology in the country. Today there are 570 KVKs in the country. Each *KVK* is headed by a Project Coordinator (PC) assisted by subject matter specialists in the disciplines of agronomy, plant protection, horticulture, animal husbandry, home science and agriculture extension. It has a demonstration farm for showing agricultural technology to the farmers. The activities of *KVK* are : (1) Training of farmers and extension workers to update their knowledge and skills; (2) Testing the location specificity of agricultural technology under various farming systems; and (3) Front line demonstration to establish production potential of improved agricultural technologies on the farmers' fields. *KVK* is also an available resource and knowledge centre of agricultural technology for institutions of public, private and voluntary sectors for improving the agricultural economy of the district.

The ICAR also started several programmes – **National Demonstration**. **Operational Research Project, Lab to Land Programme** and **Agricultural Technology Information Centre**. The scientists of research institutes of the ICAR, located in different parts of the country to educate farmers in the villages which are adopted by the institutes. Similarly, Agriculture Universities have the programmes for farmers.

In the sixties, there was a dramatic growth in agriculture production. This has transformed the national economy and has been recognized as great achievement in the field of agriculture throughout world. The word "Green Revolution" was coined to describe the achievement. At this stage it was observed that the agricultural productivity varied among different categories of farmers and regions of the country. It created small pockets of plenty leading to a wide regional disparity, and also widened economic hiatus between big farmers, the small farmers and the marginal farmers. The Government of India felt the need to reduce and eliminate this hiatus between different categories of farmers and the regions. There were two reasons for it. First, it was essential for social justice in rural society because the small farmers and the marginal farmers constitute a very large percentage of the cultivators. Second, there should be full utilization of potential of cultivated area, from national point of view for agriculture production. It was also realized that these farmers were not less progressive than the big farmers nor lacked enthusiasm for adopting modern agriculture technology. High cost of agriculture inputs, namely improved seed, chemical fertilizer and plant protection measures, were main constraints in adoption of agriculture technology for them. As a result in early seventies Small Farmers Development Agency (SFDA) and Marginal Farmers and Agricultural Labour Development Agency (MFAL) were started. These programmes were merged when Integrated Rural Development Programme (IRDP) was introduced in late seventies to improve economic level of the poor sections of the rural society. The programmes for areas with special problems or issues requiring special treatment were also started, namely: Drought Prone Area Programme, Tribal Development, Hill Area Development and Command Area Development. These programmes identified their beneficiaries and their problems, drew suitable agricultural schemes, located institutions to support and implement the programme. These programmes extend 25 per cent subsidy to small farmers to meet the capital expenditure on agriculture. The subsidy was even higher (up to 33 per cent) for marginal farmers and agricultural labor.

Agricultural Technology Management Agency (ATMA), which started in 2005-2006, was created for strengthening extension systems. It has been set up in 604 districts of 28 States and 3 Union Territories across the country and functions an autonomous body. The planning and implementation of the programme was done through bottom up approach and decentralized decision making process involving all departments associated in agricultural development in the district. The farmer is at the centre of this programme. Representation is provided to farmers at the village level as member of Farmer Interest Group, at block level as member of Farmer Advisory Committee and district level as member of ATMA Governing Board. Minimum 30 per cent of resources of the programme is utilized for women farmers. Ten per cent contribution is realized from beneficiaries toward beneficiary orientated programme. The activities under the programme include capacity building of extension workers and farmers, demonstrations on agri-culture technology, exposure visits, farmers' fairs, farmers' group mobilization and farm schools and farmers-scientist interaction.

Kisan Call Centers **(KCCs)** have been set up across the country by the Ministry of Agriculture, Government of India to provide instantaneous agriculture services to farmers over telephone. A countrywide common toll free eleven digit number 1800-180-1551 has been allotted for KCC. The number is accessible through all mobile phones and landlines of all telecom networks. Replies to the farmers' queries are given in 22 local languages. Calls are attended from 6 a. m. to 10 p.m. all seven days in a week. In order to create awareness amongst farmers about KCC programme, audio/video spots on KCC are being broadcast through All India Radio/ Doordar-shan and private Channels. A Kisan Knowledge Management System (KKMS) has also been developed to capture the details of farmers calling the KCC and their queries and also a searchable database on state wise package of practices to retrieve correct, consistent and quick reply for farmers.

Agri-Clinic and Agri-Business Centers Scheme provides exten-sion service on payment basis. The training is for one year by Natio-nal Institute for Agriculture Extension Management, Hyderabad which has identified Nodal Training Institutions in various states. These institutions provide essential support for one year.

A number of programmes in Eleventh Five Year Plan were started for agriculture development, namely, *Rashtriya Krishi Vikas Yojana*, National Food Security Mission, National Horticultural Mission, Technology Mission on Horticulture, National Watershed Development Programme for Rainfed Areas, Intensive Scheme of Oil Seeds, Pulse, Oil Palm Oil and Maize, Technology Mission on Cotton and National Bamboo Mission.

Mass media have been used to reach the farmers about agricul-ture technology. All India Radio (AIR) broadcasts programmes for the benefit of rural society through its special cell known as Farm and Home Broadcast Unit. It produces programmes on agriculture, animal husbandry, and allied subjects keeping in view the pattern of development in the target area.

Transmission of rural development programmes over television started in 1967 with the inauguration of ***Krishi Darshan* Programme** from Delhi Television Station meant for 60 villages around Delhi. Under the project government–owned television sets-were installed in the villages. The programme was of for half an hour duration twice a week (later the frequency was increased to three times a week). It was beamed from Delhi Television Station. The programme, under the title "*Krishi Darshan*", provided information on items like improved agricultural practices, success stories of agriculture innovation, news and weather reports etc. Television viewing clubs of progressive farmers were organized by the Field Unit of the project to promote the viewing of these programmes and also to encourage post viewing discussion by the farmers.

The Satellite Instructional Television Experiment (SITE) was launched on August 1, 1975 as one year programme, seeking to provide communication support and improving the social, economic and cultural life in 2400 villages across the states of Rajasthan, Bihar, Madhya Pradesh, Odhisa, Andhra Pradesh and Karnataka. The agricultural programme covered under the SITE were: (1) Dissemination of information (whenever possible with demonstration) on dryland farming, marginal farming, research findings relevant to the regions concerned, poultry and animal husbandry, package of practices recommended for important crops, timely crop operations, water management etc. (2) Information relating to organizations responsible for supply of agricultural inputs like seeds, fertilizers, implements, pesticides and other service, credit facilities, etc. (3) Pest and disease control, (4) Weather forecasts and markets. The programmes were directly received through United States ATS – F satellite situated in geostationary 36,000 kilometers in space above Victoria in Kenya.

Village Resource Centre (VRC) programme was launched by Indian Space Research Organization (ISRO), Bangalore on 19 October, 2004. Modern technology used for communication through space has made it possible for this programme (VRC) to communicate scientific knowledge to the people in villages who can benefit from it and achieve better quality of life. The Education

Satellite (EDUSAT) is utilized in this programme. EDUSAT is stationed 36000 kms from the earth. For VRC the information which is to be communicated is first sent to EDUSAT from the Earth. EDUSAT receives the signals and beams them back to ISTRAC (ISRO Telemetry Tracking and Range Centre), Bangalore. The ISTRAC then sends the signals to EDUSAT again, and from there the EDUSAT directs the signals to all VRC centres in India. The unique feature of this programme is that experts belonging to agriculture universities or research institute directly send their information, advice or opinion to villagers scattered all over the country. The people at the receiving end – the villagers-can find solution to their queries or difficulties faced by them. There are 500 VRCs in 21 State of the country.

Need

The above account shows that in post-independence era the government established institutions for educating villagers for carrying out programmes for rural development in general and agriculture development in particular. These were devised at different times considering villagers' needs, and larger national requirements and also advances in agriculture technology. The programmes have a sound conceptual framework. At present there are programmes for every section of rural society and regions in the country. The programmes have significantly raised food grain from 82 million tonnes in 1960s to 262 million tonnes in 2013-14 (Hindu, November 28, 2015) and rural prosperity. Extension workers of development departments, agricultural scientists of ICAR and agricultural universities and personnel of mass media–radio and television – all are engaged in their different ways to educate the farmers about scientific agriculture. Each of them has its own strength and weakness. However, agriculture scientists have a unique role because they are repositories of modern agriculture technology and scientific –know how of farmers' problems. As a result there is the least possibility of distortion or contamination in communication. Also, they provide directly on the spot solutions to the farmers for their problem. The prototype of this model was the work of agriculture scientists in the USA who started programmes to upgrade the knowledge of farmers pertaining to scientific farming. The ICAR has a large number of research institutes all over the country: four

National Research Institutes (deemed University), 45 research institutes, 61 All India Coordinating Research Projects, 17 National Directorates. There are 45 state agricultural universities and one Central Agricultural University. It is therefore, very useful to understand the features of extension programmes of the ICAR and review research studies to know what extension methods and technology agriculture scientists have used to educate the farmers modern farming practices, how far the farmers gained knowledge of modern farming practices and to what extent they adopted them and, most importantly to what extent agriculture productivity increased. Keeping this in view the study "Agricultural Scientists Contribution in Farmers Education about Agricultural Technology in Post-Independence India" was undertaken. The objectives of the study are presented below.

Objectives

The present study is an attempt to:

1. Discuss the genesis, importance and *modus operandi* of extension programmes of the ICAR to educate farmers in improved farming practices with the ultimate goal of accelerating agriculture development.
2. Understand the extension methods and training methodology adopted by agricultural scientists to educate the farmers about improved farming practice.
3. Assess farmers' knowledge and adoption level of improved farm practices.
4. Assess agricultural productivity enhancement as a result of the programmes.
5. Identify the operational problems in farmers' education regarding improved farming practices.
6. Suggest extension strategy for promoting farmers' education about improved farming practices

Method

The author collected research papers on extension programmes from research journals, books, government reports and newspapers. The author was himself involved for a long time in extension programmes of the ICAR at Central Arid Zone Research, Jodhpur as Principal Scientist, the experience of which also provided a great help in the study.

Scope

The study deals with contribution of agricultural scientists in farmers' education about improved agriculture practices in post-Independence, India. The work done for the farmer education by the agriculture scientists is invaluable for the country, even for people of the world over. The policy makers, administrators and extension workers of the countries have gained from the experience and can provide valuable input for strengthening farmers' education programme about scientific know-how of agriculture. This will ultimately bring agriculture advancement which is concern for all the countries.

The book has been made simple with accent on meaningfulness rather than sophistication.

Limitation

Research studies published from research journals and reports were reviewed. There were a few reviews from unpublished theses. The study was conducted by the author himself and due to financial constraints it was not possible to collect information from more theses and to visit various agriculture universities.

REFERENCES

Nataraju, M.S. and Lovely, P.G., "Role of Farm Women in Farm and Home Activities", Indian Journal of Adult Education, 50(4): 64-67, 1989.

Saikia, A., "Effects of Cropping Pattern on Employment of Female. A Study in Sibsagar and Jorhat Districts Assam", Indian Journal of Agricultural Economics, 60(3): 264, 1985.

Tiwari, P., Waris, A and Singhal, S., "Rural Women's Participation in Agriculture and Livestock Activities in Arid Area of Rajasthan", Current Agriculture, 20 (1 & 2): 91-95, 1996.

■■■■

CHAPTER - 2

National Demonstration

ORIGIN

In early sixties, there was a great advancement in scientific know-how in the area of agriculture technology. The scientists developed high yielding varieties (HYV) and their cultivation method. The high yielding varieties were (1) Lerma Rojo and Sonora 64 (Wheat), (2) Taichung Native 1 (Paddy), (3) Ganga 1, Ganga 101, Ranjit and Deccan (Maize), (4) C.S.H.1 and C.S.H. 2 (Sorghum) and (5) H.B. 1 (Pearl millet). There was dramatic increase in the yield of the high yielding varieties. Over ten tonnes of grain was harvested in a hectare from HYVs of wheat and maize in year at I.A.R.I., New Delhi. Thus there was a great opportunity to raise food production in the country. It was essential that new techniques of HYV were adopted by farmers to obtain the full potentiality of the yield. However, extension workers and farmers did not possess the required knowledge and skill in the application of new techniques of agriculture. In 1964, **Dr. M.S. Swaminathan** (then Director, I.A.R.I., New Delhi) proposed to a Panel of Scientists as its member, that practical demonstration (**National Demonstration**) should be done by scientists for farmers at their fields. He said that National Demonstration would differ with demonstration conducted by extension agencies in three respects. Firstly, there would be a specific yield target such as 3.5 tonnes per hectare in wheat and there would be no separate control plot near the demonstration field. The idea behind this principle was that the entire living memory of the farmers concerning the yield potential of the crops as well as the entire block in which the demonstration has been laid should serve as the control. Secondly, the area of the demonstration plot should be one hectare so that the feasibility of raising a good crop could be strikingly and unquestionably demonstrated. Third, the farmers in whose plots the demonstration is done should be actual cultivators with small holding, so that the high yields obtained are not attributed to the effects of affluence. The Panel of Scientists which had Shri D.K. Ramaiah as its Chairman and Dr. B. Pal as its convener was constituted by the Union Ministry of Food and Agriculture, New Delhi. The programme was approved by the Union Ministry of Food and Agriculture, New Delhi and was started in 1965.

While inaugurating this programme the Union Minister for Food and Agriculture remarked: "Unless scientists can demonstrate what they advocate, their advice may not be heeded. Since they have not had much opportunity to do this in the past, we are initiating during 1965 a 'National Agriculture Year Programme' , an important part of which will be to stud the country with excellent demon-stration plots showing the yields that can be easily obtained in all our major crop plants, provided the available scientific know-how is put to use. These demonstrations will be laid out in farmers' fields by the staff and students of all our agricultural institutions and colleges. They will thus help in dispelling the views expressed on the one hand by some politicians and administrators that our scientists have little contact with the farming community and on the other, by our senior scientists that what is applied research of great value is constantly converted into 'ivory tower' research in our country by not making proper arrangements to exploit the research findings".

This programme of demonstration, therefore, has been on unique phenomenon in the history of Indian agriculture. Since the scientists were entrusted with the direct responsibility of demonstrating in the fields of farmers, that the results obtained by them in experiment stations are of wide applicability and utility.

MAIN FEATURES OF NATIONAL DEMONSTRATION

1. There is specific minimum yield target, but there is no specific control plot near the demonstration. The idea behind this principle is that the entire living memory of the farmers concerning the yield potential of the crop as well as the entire block in which the demonstration has been laid should serve as control.

2. The objective of the national demonstration is to introduce the concept of the productivity per unit area time, in which the unit area is kept one acre/hectare and the unit of time one year.
3. The target of multiple cropping with 9/11 tonnes per hectare minimum grain yield in two or three crops respectively was fixed. Thus the multiple cropping which is a nature's gift to tropical area has been made the key point of national demonstration.
4. These demonstrations are not in any way intended to supplement the exiting departmental demonstrations meant for creating awareness for a particular improved practice laid out by the extension workers. The uniqueness lies in the fact that the national demonstrations are laid out by the researcher who will get an opportunity to demonstrate in farmers' fields what science can do to transform Indian agriculture.
5. The layout of these demonstrations is the responsibility of the four subject matter specialists who are experts in the disciplines of agronomy, soil science, plant protection and agricultural engineering.
6. The farmers in whose plots the demonstration is laid are to be actual cultivators with small holdings, so that the high yields obtained are not attributed to the effects of affluence. Scientist of research stations/agricultural universities have to demonstrate convincingly to farmers that the results which are obtained at research stations could be repeated on the farmers' fields as well.
7. In place of generalized package of practices, the approach in these demonstrations is that the operation in various stages of crop production are to be determined by specific site factors such as slope, depth, texture, drainage characteristics, presence or absence of soluble salts etc. Thus every demonstration is unique entity which shows the practices to be followed and yield to be expected on specific sites.
8. Very intensive cropping system techniques of soil and water management, including use of machinery and plant

protection are demonstrated. Thus, these demonstrations serve as ***pace setters*** which are to be the first line demonstrations and are to be emulated by local extension workers and farmers.

9. These demonstrations are to serve as 'class rooms' where farmers are taught the appropriate techniques to be followed in crop management.
10. Even new problems requiring solutions will emerge from these demonstrations. They will be fed to research organizations for their solutions.
11. These demonstrations provide an active bridge between research, farmers training and the food production programmer of the country. Hence these demonstrations are termed as ***national demonstrations***.

DEMONSTRATION

It is essential to be clear about meaning, method of conducting demonstration, advantage and limitation of demonstration

The demonstration is the foundation of village extension work. Demonstration is based on the idea of showing rather than writing or talking. It presents a proven improved practice in terms of its application to a specific situation. There are two types of demonstrations: (1) Method Demonstration and (2) Result Demonstration.

1. Method Demonstration

A method demonstration is a short –time demonstration given before a group to show how to carry out an entirely new practice or an old practice in a better and modified way. This method, however, is not concerned with proving the worth of a practice but with how to do something. It is definitely not an experiment or a trial, but a teaching effort.

The purpose of this demonstration is to effectively impress upon people that a particular recommended practice is also able proposition in their own situation. It teaches a new skill.

The demonstration units are told that they should :

1. Be at the spot early to check up equipment and material.

2. Make physical arrangements so that all participants can have a good look at the demonstration and take part in the discussion.
3. Conduct demonstration step by step, encouraging questions at each stage.
4. Give opportunities to individuals to practice the skill.
5. Distribute bulletins, leaflets, etc., related to the demonstration.
6. Give the names of the participants and list those who contemplate the adoption of the practice. This helps in the follow-up and increases the number of persons desiring the change.
7. Publish a news story on the demonstration.
8. Entrust the leaders with the task of watching the adoption of the new practice.
9. Select topics based on the needs of people.
10. Demonstration should be timely.
11. Give good advance publicity to build up interest and secure wide participation.
12. Use equipment that is easily available to rural people.
13. Clear doubts, but avoid arguments.
14. Appreciate the methods already being used by the participants.

Some of the specific jobs that could bc taught through this method are – nursery bed, line sowing, improved implements, application and placement of fertilizers, seed selection, eradication of ticks and mites in cattle, etc.

Advantages

1. Effective in teaching skills to farmers,
2. Stimulates action,
3. Builds confidence,
4. Serves publicity purpose,

5. Increases acquaintance of extension workers with local people, and introduces change of a practice at low cost.

Limitations

1. Not suited to all situations.
2. Needs great deal of preparation, both of equipment and skill on the part of the extension worker.
3. Causes a set-back to whole programme if improperly coordinated

2. Result Demonstration

A result demonstration is a way of showing people the value of a new practice. Such demonstrations require a substantial period of time. Comparisons are usually necessary and records are essential. The result demonstration may be for a single recommended practice or a series of practices that come in sequence with respect to a problem.

The result demonstration is preferred: 1. To prove the worth of a new practice. 2. To show that the recommendations are locally applicable.

Planning for such demonstration involves various activities like

1. Discussion with the people and find out their interest.
2. Consultation specialists.
3. Outlining a definite plan and its documentation.
4. Selecting responsive farmers in consultation with the local people.
5. Discussion and planning with selected farmers.
6. Announcing widely.

Conducting the demonstration would require the conducting unit to

1. Get all the materials ready.
2. Start the demonstration in the presence of the villagers.
3. Mark demonstration plots with large signs if it is field demonstration.

4. Assist the farmer, on whose field the demonstration is conducted, in keeping proper records.
5. Make a calendar of operations.
6. Visit the demonstration site regularly for supervision and guidance.
7. Give publicity to the demonstration and the farmer at suitable stages, preferably through other farmers' visits.
8. Conduct tours to successful demonstrations at proper times.
9. Let the selected farmer tell people about the demonstration, when possible.
10. Summarise, record and establish proof the practice.
11. Give wide publicity to the results.
12. Enlist other farmers to adopt the new practice.

Suggestions

1. Limit the scope of the demonstration to prove the advocated practice rather than finding out a new truth.
2. Have a comparison or check plot.
3. Use local illustrations of good practices rather than result demonstration, whenever possible in order to save time and extension effort.
4. It is better to conduct one good demonstration rather than a number of half- hearted attempts.
5. Let the benefit of the demonstration reach as many local people as possible through conducted tours, publicity, etc.
6. Select cooperative farmers in different areas to undertake.
7. Demonstrations, avoiding sticking to the same farmers for several demonstrations.
8. Select places for demonstrations which are frequented by the villagers.
9. Use the results of successful demonstrations in meetings, individual contacts, and newsletters, etc.
10. In case of failure analyse the causes carefully.

Such demonstration is a preferred method in cases of improved seeds, fertilizer application, plant protection measures, castration of scrub bulls, artificial insemination, etc.

Advantages

1. Helps in introducing a beneficial new practice.
2. Convinces the hesitating farmers and workers as they see themselves how better yield or result has been attained.
3. Provides factual data.
4. Gives the extension worker experience that helps them to recommend the practice with greater conviction.
5. Develops local leadership.
6. Convinces farmers that the extension workers are equipped with practical knowledge.

Limitations

1. Finding a suitable farmer willing to keep records is often difficult.
2. Results may be vitiated by uncontrollable factors like weather.
3. An unsuccessful demonstration may create strong unfavorable conditions for other extension activities.
4. It is not suitable for all advocated practices.
5. Requires elaborate and through preparation; its cost is very high.
6. Creates rivalries sometimes.

■■■■

CHAPTER - 3

Research Studies on National Demonstration

The ***National Demonstration Programme*** described earlier has been a unique intervention by the government as the agricultural scientists who are repositories of scientific knowledge of agriculture were involved for the first time in farmers' education and transmission of scientific knows to the farmers. A team of scientists comprising experts in the departments of agronomy, plant protection, extension science and water management was formed in each research institution of ICAR, New Delhi. The agricultural universities are also responsible to conduct demonstration of improved agriculture practices at the farmers' fields. The potentiality of such attempts is large. However, what has been the actual impact of the programme on crop productivity and behavioral changes among the farmers and how the programme was implemented in the villages? The results have been assessed by the agricultural scientists. The following paragraphs look at the findings of the agriculture scientists are discussed below.

Section 1
IMPACT OF NATIONAL DEMONSTRATION ON CROP YIELD AT FARMERS' FIELD

The earliest evaluation in the year 1966 of National Demonstration Programme was reported by then Director **Swaminathan, M.S.** IARI, New Delhi. 365 crop demonstrations were started in the first year of the programme in 1965 in different parts of the country. National Demonstration produced very positive results. There was very high yield of crops in demonstration plots: paddy 4,684 kg/ha, maize 4,007 kg/ha, sorghum 2,626 kg/ha, pearl millet 2,365 kg/ha and finger millet 3000 kg/ha (Table 3.1). In nearly all the demonstrations, the yields far exceeded the stipulated targets. In 1965, the highest yield 8695 kg/ha of paddy was obtained at the field of Shri M. Badraiah, Gandhi Masaipat village, Yellareddy Block, Nizamabad district in Andhra Pradesh. Other good yields were: 5906 kg/ha at Ambasamudram in Tamil Nadu with variety A.S.D-1, 5777 kg/ha at Amritsar in Punjab with H.B 27 and 5461 kg/ha at Deoghar in Bihar with BR 34. Such good yields obtained throughout the country during an unusually dry year were indicative of the great potential of Indian agriculture.

Table 3.1. Results of National Demonstrations, Kharif 1965

Items	**Rice**	**Maize**	**Sor-ghum**	**Pearl millet**	**Finger millet**
Number of demonstrations proposed	153	69	74	75	10
Number of demonstrations from which results are available	100	60	49	57	3
Average targeted yield (kg/ha)	3,791	4,732	2,542	2,365	3,000
Average yield secured (kg/ha)	4,684	4,077	2,626	2,940	4,289
Percentage increase or decrease over target	+16	-16	+3	+25	+43
Highest yield recorded (kg/ha)	8,695	6,192	6,175	5,175	5,202
Present All- India average yield (kg/ha)	1,610	993	545	381	791

Source: M.S. Swaminathan, National Demonstration, Indian Farming, XVI, 6, pp.68-70, 1966.

Dr. Swaminathan said on this achievement : "The yield obtained in the first year of these demonstrations is ample evidence of the enthusiasm and dedication of the scientists who undertook this task, although this has to be as an additional item of work in the midst of a heavy load of scientific and teaching work." Further he said: "The National Demonstration organized during 1965 served as window into the world of plenty and prosperity awaiting farming community provided science is introduced into farming. This helped dispel doubts concerning the feasibility of getting high yields in cultivator's fields. They generated a hunger for seed and fertilizer. This in turn destroyed the concept that Indian farmer is averse to change. Our farmers proved to have the same reaction to change as everywhere if the return is good, the will to adopt a new technology is high."

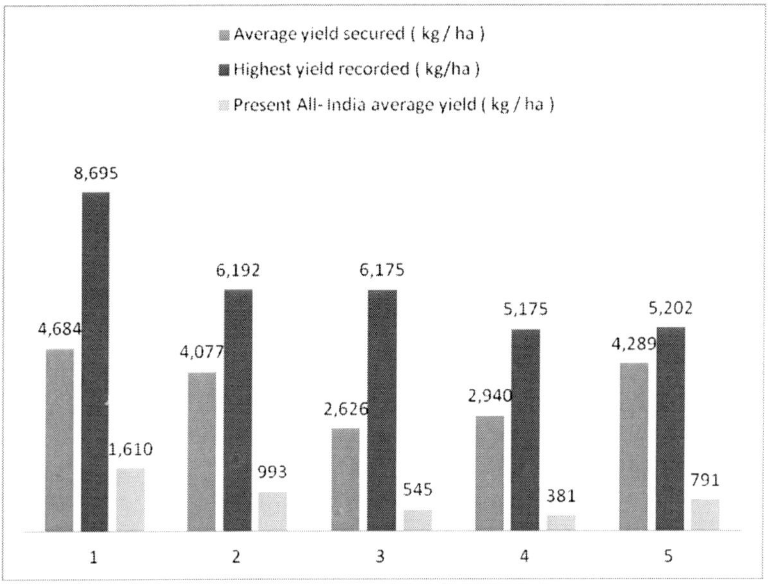

Fig. 3.1. Results of National Demonstrations, Kharif 1965

J.S. Kanwar (1971) on the analysis of five years data of crop yield of demonstration reported that a farmer could produce four to five tones more food per hectare as compared to that obtained from to the traditional methods. The farmers having irrigation all the year round produced grain yield as high as 23 to 24 tonnes per hectare per annum through multiple cropping. It showed that with the proper use of technique a farmer was able to produce more food per unit area and feed many more mouths per hectare than it was possible with the traditional system of agriculture. A large percentage of farmers obtained more than the targeted yield per hectare of wheat and paddy; but so far as maize, sorghum and pearl millet hardly 30 to 50 per cent farmers were able to exceed the targets. Two crops rotation showed that where maize, pearl millet and sorghum are one of the crops; the percentage of cases exceeding the targeted yield of nine tonnes per hectare was low. It showed poor results in the case of maize, pearl millet and sorghum. The scientists needed to find better varieties, and extension workers needed to make more efforts for disease and pest control and water management for these crops.

Table 3.2. Highest Yield (Q/ha) obtained and inputs used during 1969-70

Particular	**Paddy-Wheat-Green gram**	**Paddy-Paddy**	**Paddy-Wheat**	**Maize-Wheat**	**Paddy-Paddy-Wheat**
State and District	Delhi (Delhi)	Shimoga (Mysore)	Bundi (Raj)	Chindwara (M. P.)	Coimbatore (Tamil Nadu)
Yield (q/ha)	Paddy-83, Wheat-63-Green gram-5	Paddy-105, Paddy-103	Paddy-108, Wheat -48	Maize-74, wheat-60	Paddy-47, Paddy-53, Paddy -86
Total yield	151	208	156	134	186
Fertilizer used					
F.Y.M. (Cart Load./ha)	-	25	25	37.5	74.5 (T./ha)
N (kg/ha)	250	337	248	270	383
P_2O_5 (kg/ha)	160	160	95	110	201
K_2O (kg/ha)	67	107	68	75	184
Irrigation	12	Flow continuous	11	11	Once in two days
Cost of inputs (Rs)	1308	3003	1554	1340	3325
Gross return excluding the cost of fertilizer & pesticide	8755	7422	8191	8482	5941

Source: Kanwar, J.S. 1971, National Demonstration: Retrospect and Prospect, Indian Farming, XXI, 6, pp.13-17, 1971.

The best yields obtained by the farmer in some selected crop rotation in 1969-70 are given in the Table 3.2. It showed that the farmer who obtained the highest yield in different rotation also obtained the maximum gross return after paying the cost of inputs purchased from the market. Thus his net return was maximized. The success in obtaining higher yield provided remunerative employment for the farmer and his family. It increased the number of operations and hence the need for employing more labour. Thus it opened up numerous new avenues of employment for skilled and unskilled labour.

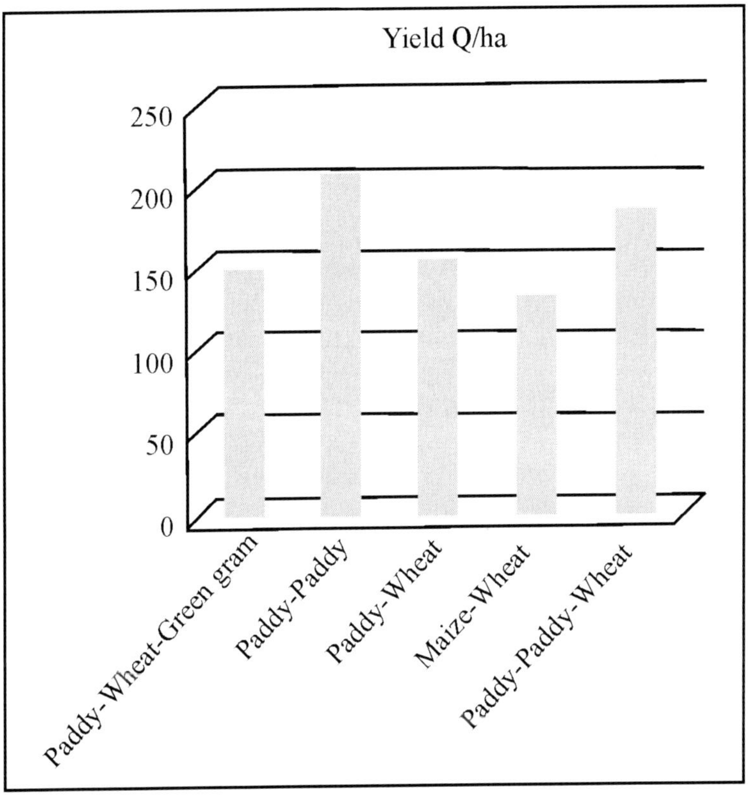

Fig. 3.2. Highest Yield (Q/ha) obtained and inputs used during 1969-70

The study of **Malhotra and Katyal** (1986) on national demonstration undertaken by Haryana Agricultural University, Hissar in Bhiwani district from 180-81 to 1984-85 on improved agricultural practices of crops shows similar results. The yield of crops in demonstration in irrigated condition was 60 to 249 per cent higher than the state average yield. Paddy included in 38 demonstrations gave an average yield of 60.3 quintals per hectare in national demonstrations which is 60 per cent higher than the state average yield of 37.8 quintals per hectare. 79 demonstrations on pearl millet were conducted and the mean yield obtained was 21.3 quintals per hectare which was 249 per cent higher over the state average yield of 6.1 quintals per hectare. One hundred and twenty eight demonstrations on wheat were laid out. The mean yield of 37.2 quintals per hectare was 51 per cent higher than the state average of 24.7 quintals per hectare. Similarly the mean yield of chick pea (13.3) and cluster bean (13.00) of demonstration plots was 183 and 105 per cent more respectively than the state average yield of these crops. A comparison of the highest yield (q/ha) obtained under the national demonstrations (Paddy 80, pearl millet 40, wheat 64.0, chickpea 20.0, greengram 12.5, redgram 25.0, clusterbean 25.0, mustard 25.0) showed that the gap is really large. Even if the highest yields obtained in national demonstrations plots are ignored and the national demonstrations mean yield are taken as yardstick, there is a great possibility to raise the crops yield provided the farmers adopt the improved high-yielding varieties and related package of practices on their farms. The results of demonstrations conducted under rainfed conditions also showed high yield (q/ha) of crops: pearl millet 14.8, cluster bean 11.0, green gram 12.0 and chick pea 15.0.

Table 3.3. Performance of major crops (q/ha) in National Demonstration in Haryana 1980-81 to 1984-85

Crops	National Demonstration highest yield	National Demonstration average yield	State average yield	Gap (%)
Paddy	80	60.3	37.8	60
Pearl millet	40	21.3	6.1	249
Wheat	64.0	37.2	24.7	51
Chick pea	20.0	13.3	4.7	183
Green gram	12.5	10.0	-	-
Red gram	25.0	20.3	-	-
Mustard	25.0	18.9	-	-

Source: Malhotra, O.P. and Katyal, S.K., National Demonstrations in Haryana show the way to progress, 1986, Indian Farming, XXXVI, 9, pp. 35-36, 1986.

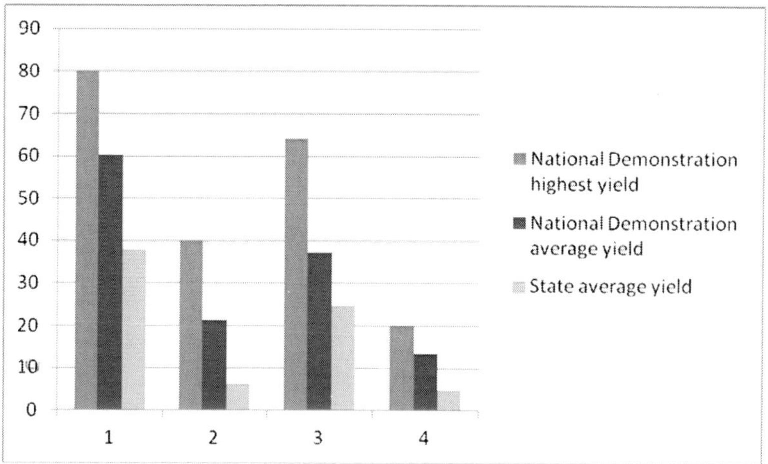

Fig. 3.3. Performance of major crops (q / ha) in National Demonstration in Haryana 1980-81to 1984-85.

Ananthanarayana *et al.* (1990) reported the impact of demonstrations conducted at farmers' fields in three *taluks* of Chitradurga district from 1983 to 1988. The yield of cereals and oilseed crops was increased two to three folds as compared to state yield. Similarly a substantial increase in yield took place after the use of soil improvement techniques in the problematic area. It was also clear that not only the yields were higher in National Demonstration Plots but return on rupee invested was also higher. Thus the technology transferred in National Demonstration Plots is economically viable.

Table 3.4. Yield gap in major crops in National Demonstration Project in Chitradurga district (1983-84 to 1987-88) Q/ha

Type of farming	Crops	National demonstration Highest yield	National Demonstration average yield	State average yield	Gap in yield (%)
Irrigated	Rice	90	68	23	195
	Finger millet	53	42	13	223
	Maize	80	55	16	243
	Groundnut	32	28	10	180
	Sunflower	32	22	8	175

Type of farming	Crops	National demonstration Highest yield	National Demonstration average yield	State average yield	Gap in yield (%)
Rainfed	Sunflower	15	9	3	200
	Groundnut	15	12.5	5	150

Source : Ananthanarayana *et al.*, National Demonstrations show the way to increase crop yields in Chitradurga district, Indian Farming, XL, 6, pp. 15-16, 1990.

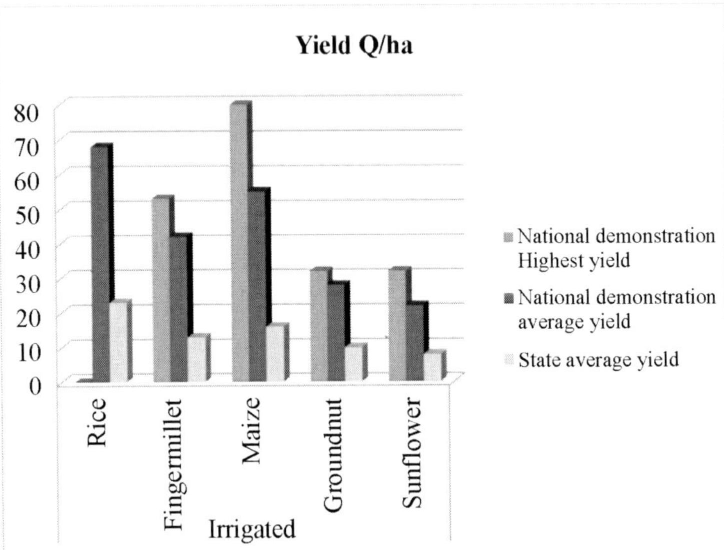

Fig. 3.4. Yield gap in major crops in National Demonstration Project in Chitradurga district (1983-84 to1987-88) Q/ ha

Table 3.5. Results of special demonstration

Crop	Variety	Treatment	Yield (q/ha)
Kharif			
Finger millet	Indaf 9	Gypsum+ $ZnSO_4$	32.50
Finger millet	Indaf 9	Control	22.50
Summer			
Paddy	IR 20	Gypsum+ $ZnSO_4$	37.50
Paddy	IR 20	Control	22.50

Cost –benefit ratio in National Demonstration Project for major crop

Crop	Year	Ratio	Year	Ratio
Rice	1983-84	2.2	1987-88	3.3
Finger millet	1983-84	2.2	1987-88	3.3
Groundnut	1983-84	2.0	1987-88	2.7

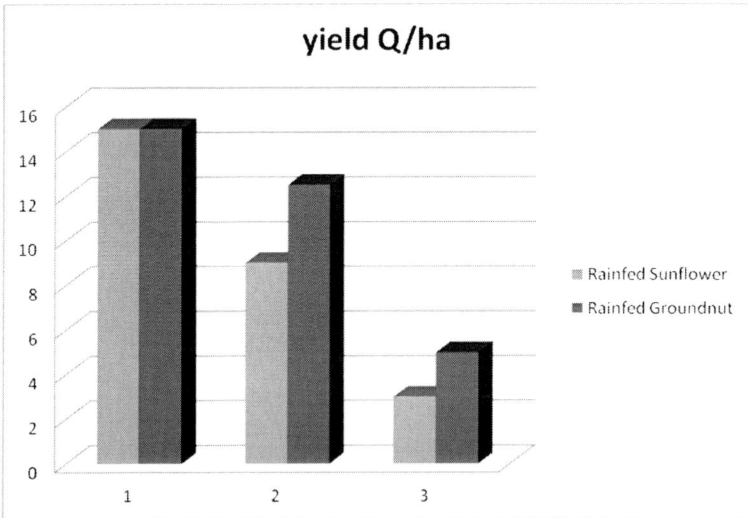

Fig. 3.5. Yield gap in major crops in National Demonstration Project in Chitradurga district (1983-84 to1987-88) Q/ ha

Goydani *et al.* (1988) studied impact of demonstrations conducted by the Jawaharlal Nehru Krishi Vishwa Vidyalaya's Regional Agricultural Research Station, Chhindwara in Madhya Pradesh. They found a wide gap ranging 64 to 199 per cent and 76 to 287 per cent in the yields of various crops as compared to district and state average yields respectively. Soybean gave an average yield of 16.88 quintals per hectare which was 52 per cent and 11 per cent higher than the district average and state average yields of 11.09 and 8.00 quintals per hectare respectively. The mean yield of gram was 10.48 quintals per hectare which was 121.56 per cent and 76.72 per cent higher than the district and state average yields of 4.73 and 5.93 quintals per hectare respectively. The mean yield of 32.24 quintals per hectare of wheat was 199.62 and 270.57 per cent higher than district and state averages of 10.76 and 8.70 quintals per hectare respectively. The average yield of 35 quintals per hectare of maize was 110.33 per cent and 287.60 per cent higher than district and state average of 16.64 and 9.03 quintals per hectare respectively. He concluded on the basis of the performance of all crops during the period from 1983-84 to 1986-87 that the productivity of all crops could be increased considerably provided the farmers adopt the improved high-yielding varieties associated with proper package of practices on their fields.

Table 3.6. Yield performance yield (q/ha) of major crops under National Demonstration Project in Chhindwara District of Madhya Pradesh from 1983-84 to 1986-87

Crop	Highest yield in national demonstration	Average yield in national demonstration	State yield	Gap (%)	District yield	Gap (%)
Soybean	24.59	16.88	8.00	111	11.09	52.20
Gram	17.59	10.48	5.93	76.72	4.73	121.56
Wheat	48.50	32.29	8.70	270.57	10.76	199.62
Maize	54.86	35.00	9.03	287.60	16.64	110.33

Source: Goydani *et al.*, National Demonstrations in Madhya Pradesh show the way to progress. Indian Farming, XXXVII, 9, pp.31-32, 1088.

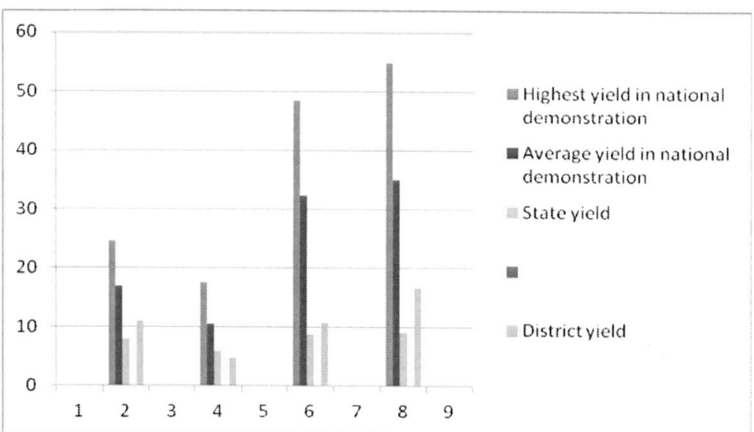

Fig. 3.6. Yield performance yield (q/ha) of major crops under National Demonstration Project in Chhindwara District of Madhya Pradesh from 1983-84 to 1986-87

Dutt *et al.* (1983) studied yield difference between National Demonstration plots and state averages got from general farmers from the year 1973-74 to 1981-82 in the districts of West Tripura. He reported that maximum potentiality for increasing the yield of rice was 281 per cent as compared to the state average yield and 247 per cent as compared to the district (west Tripura yield). There exists considerable potentiality 18 per cent to 124 per cent of increasing the yield of potato over the state average. It is between 3 per cent to 50 per cent for wheat and between 82 per cent to 112 per cent for jute.

In more than 70 per cent demonstrations, the yield was more than the targeted yield of 9 tonnes per hectare and 11 tonnes per hectare for two crop and three crop demonstration respectively. Thus it is feasible to obtain 9 to 11 tonnes per hectare yield by following suitable crop rotation and management.

Table 3.7. Yield difference between national demonstration and State average and district in Tripura from 1973-74 to 1981-82

Crop	National demonstration yield (q/ha)	State yield (q/ha)	Gap%	District yield (q/ha)	Gap%
Rice	33.33 to 47.29	1.91 to 13.56	86 to 247	11.73-19.89	96-247
Potato	65.62 to 202.50	68.00to 137.90	18 to 124	48. 36 177.00	10 to 125
Wheat	24.05 to 25.60	20.00 to 23.19	3 to 50	19.97 to 26.66	9 to 50
Jute	25.62 to 30.00	12.06 to 16.43	82 to 116	13.04 to 23.16	29 to 105

Source: Dutt *et al.*, National Demonstrations in Tripura, Indian Farming XXXIII, 9, pp. 45-47, 1983.

The study by Singh *et al.* (1994) of the National Demonstration Programme implemented by the Govind Ballabh Pant University of Agriculture and Technology, Pantnagar in Shahjahanpur district of Uttar Pradesh shows the positive impact of the programme result. The study examined 394 national demonstrations on rice, wheat and groundnut crops which were conducted from 1985-86 to 1989-90. The yields of rice under National Demonstration were higher by 50-100 per cent over the corresponding average village and district yields and it was 102 to 175 per cent higher as compared to that of the state average. The yields of rice in the demonstration plots ranged from 4.75 to 7.99 t/ha the average being 5.33 t/ha during the study period. The yields of wheat in the National Demonstrations plots were higher by 1.5 to 2 times as compared to the local average and up to 2.5 times in comparison to those of the district and state averages. In National Demonstration Plots yields were in the range of 3.74 to 5.60 t/ha, average being 5.06 t/ha, against the average of about 2.1-2.2 t/ha of district and about 2.0 t/ha of the state. The yields of groundnut on demonstration plots were higher *vis-a-vis* those of local, district, state averages; it was 23.5 times higher than the local average and 2 to 4 times as compared to the district and state average yields. The maximum yields of the demonstration plots were 3.05 and 1.65 t/ha respectively with 2.29 t/ha as average yield for the 4years during which the demonstrations were laid out.

The average cost of cultivation of groundnut which ranged from Rs. 3,058 to 4,366 during the period, was about Rs 3,828/ha. The net profit per rupee invested, however showed a declining trend; it was Rs 2.58 in 1986-87, but only Rs 1.70 in 1989-90. The decline was mainly due to increase in the cost of cultivation The net profit per hectare from rice and wheat crops were Rs 5,000 and Rs 5,400 /ha respectively, whereas, it was Rs 7,500 in case of groundnut which gave the highest Rs 1.63 net return. Though the cost of cultivation per hectare was higher in national demonstration plots owing to the adoption of the package of new agriculture technology, the net return per rupee invested was higher.

Table 3.8. Maximum, minimum and average yields (t/ha) obtained in national Demonstration with village, district and state average yields in different years

Year	Rice (Yield)					Wheat (Yield)						Groundnut (yield)					
	ND 1	ND 2	Vill-age	Dist-rict	Stat e	ND 1	ND 2	ND 3	Vill-age	Dist-rict	Stat e	ND 1	ND 2	ND 3	Vill-age	Dis-trict	State
1985 -86	7.00	5.00	5.08	2.80	NA	5.40	3.74	4.73	2.20	NA	2.00	2.50	1.87	2.19	.80	.59	.88
1986 -87	6.25	4.50	5.64	3.00	2.70	5.26	4.65	5.12	2.50	2.14	1.93	3.05	1.65	2.32	1.04	.58	.90
1987 -88	5.95	4.75	5.25	3.20	2.80	5.40	4.70	5.04	2.80	2.18	1.19	2.90	1.75	2.28	1.15	.57	.61
1988 -89	6.30	5.20	5.59	4.00	3.60	5.60	4.80	5.44	3.00	2.46	2.25	2.80	1.90	2.36	1.45	1.19	.96
1989 -90	6.50	4.95	5.49	NA	NA	5.20	4.54	4.95	3.20	NA	2.05						

Note: ND1 – National Maximum, ND2- National Minimum and ND 3- National Average

An excellent grain yield of finger millet crop of 5202 kg/ha at G. Ramaswami' s plot and 3915 kg/ha at V. l. B Balakrishnan's in Coimbatore was obtained. The yield of demonstration plots exceeded the target yield of 3000 kg/ha. The net profit was Rs 3,103 and Rs 2128.80 respectively. The farmers present at harvesting were very much impressed by the crop stand and high yield of finger millet crop. The extension staff was so much impressed with the performance of finger millet- Co.7 that the entire produce of demonstration plots was purchased for distribution as primary seed. (Shanmugasundaram *et al.,* 1966).

Table 3.9. Yield of irrigated finger millet demonstrations at farmers' fields

Particulars	**Vedpatti**	**Karadi Bhavi**
Grain (kg/ha)	5,202	3,915
Straw (Green) kg/ha	34,780	27,170
Cost of cultivation (Rs)	904.60	917.85
Value of produce		
Grain (Rs)	2,965.12	2,231.55
Straw (Rs)	1,042.50	815.10
Total (Rs)	4,007.64	3,046.65
Net Profit (Rs)	3,103.00	2,128.80

Source: Shanmugasundaram *et al.*, National Demonstration impresses Coimbatore ragi growers, Indian Farming, XVI, 4, pp.7-8, 1966.

In 1976, a phenomenal yield of 7711 kg/ha was obtained at the demonstrations of multiple cropping of finger millet (Variety K 2) and pearl millet (Variety H B-1) at the field of Gursamy Nadar at Illuppaiyurani village in Tamil Nadu. The net income of Rs. 3,896 per hectare was obtained by adopting the two crop sequence within a period of eight months. The yield of demonstration plot exceeded the stipulated target (Subbaiah *et al.,* 1969). A phenomenally high yield (10863 kg/ha) of rice was obtained at Baidhar Sahu's field in Fakirapada in Cuttack in 1967. Such high yield was obtained with an investment of Rs 1250 and resulting in a net profit Rs. 3200 per hectare from a single crop of rice during a period of four months. By adoption of three crops (rice-potato and rice) in a year Shri Mahapatra a resident of Naharkanta in Cuttack got a very high yield 32470 kg per hectare and return of Rs. 5698 (Mahapatra, 1969)

Table 3.10. Impact of national demonstration at the field of Shri M. Gurusamy Nadar.

Particulars	**K.2 Finger millet**	**HB-1 Pearl millet**	**Total for Finger millet and Pearl millet**
EXPENDITURE			
Cost of cultivation for one hectare (Rs)	1388.70	1093.79	2482.49
RECIPTS			
Yield of finger millet and pearl millet (kg)	3645	4066	7711
Yield of finger millet and pearl millet (Kg)	9375	7410	16785
Value of finger millet grain at 79 paise and Pearl millet grain at 75 paise per kg (Rs)	2879.55	3049.50	5929.05
Value of finger millet straw at Rs 40 and pearl millet at Rs 10/tone (Rs)	375.00	74.10	449.10
Total receipt: (Rs)	3254	3123.60	6378.15
NET PROFIT : (RS)	1865.85	2029.81	3895.66
Production of grain per day (kg/ha)	30.4	47.8	32.0

Source: Subbaiah *et al.*, National Demonstration of finger millet and pearl millet impresses farmers, Indian Farming, XIX, 6, pp. 29-31, 1969.

By adoption of three crops (rice-potato and rice) in a year Shri Mahapatra a resident of Naharkanta in Cuttack got a very high yield 32470 kg per hectare and return of Rs. 5698 (Mahapatra, 1969).

Crop rotations was found to be more promising in various regions of the country as seen from the analysis of the results of National Demonstration Programme. On the basis of the results obtained from the National Demonstration Projects, some of the crop rotations were found to be more promising in various regions of the country (Sharma, *et al.*, 1982). These are discussed below.

Two crop rotation

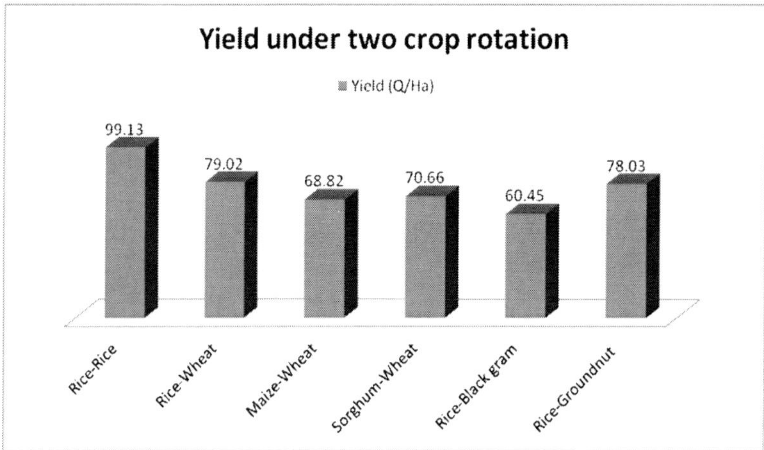

1. Rice–Rice–Rotation: It proved successful in erstwhile Andhra Pradesh (Chittoor, Hyderabad and Guntur), Karnataka (Raichur and Chikkmagalur) and Orissa (Ganjam and Koraput). The mean yields ranged from 108.82 q/ha to 124. 46 q/ha for two crops of rice. In 81 per cent demonstrations with this rotation, the yield exceeded the target of 90 q/ha. The mean yield obtained in these states was 99.13 q/ ha.
2. Rice–Wheat Rotation: This rotation was followed in various states in the country and mean yield obtained from these demonstrations was 79.02 q/ha. This rotation was found more promising in Bihar, Punjab and Uttar Pradesh.
3. Maize – Wheat Rotation: The demonstrations were conducted with maize–wheat rotation in Himachal Pradesh, Karnataka, Punjab, Rajasthan and Uttar Pradesh. The mean yield was 68.82 q/ha. This rotation was popular in Himachal Pradesh, Rajasthan and Uttar Pradesh.
4. Sorghum - Wheat Rotation: The all India mean yield from demonstrations conducted with sorghum–wheat rotation was 70.66 q/ha. The demonstrations were mainly conducted in Chandrapur and Yeotmal (Maharashtra).
5. Pearl-millet –Wheat Rotation: The demonstrations were conducted in Haryana, Maharashtra, Rajasthan and Uttar

Pradesh with this rotation. The performance of this rotation was found promising in in Jaipur.

6. Rice – Blackgram Rotation: This rotation was followed in Andhra Pradesh, Orissa and Tamil Nadu. The average yield of grain was 54.02 q/ha and that of blackgram 6.43 q/ ha.

7. Rice – Groundnut rotation : The mean yield from the demonstrations conducted with rice –groundnut in Andhra Pradesh, Karnataka, Kerala, Odisha, Tamil Nadu and Goa was 55.86 q/ha of grain and 22.17 q/ha of pods

Three crop rotation

1. Rice-Rice-Rice Rotation: Three crops of rice in the same field was followed in Pondicherry and the mean yield obtained was 140.62 q/ha.

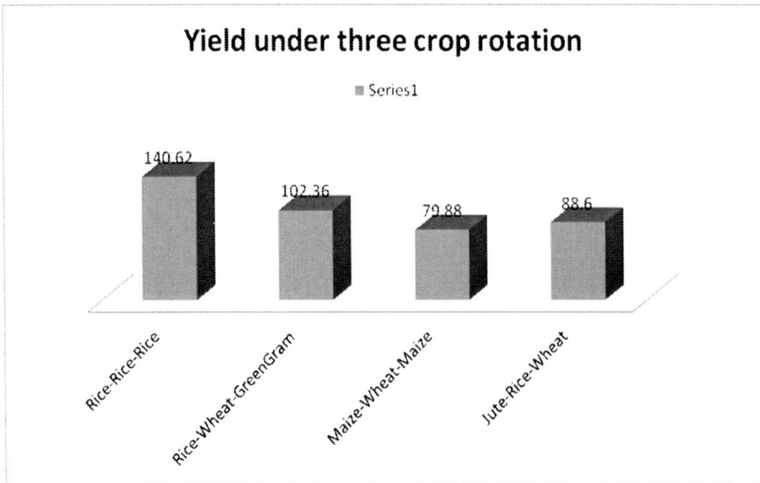

2. Rice –Wheat – Green gram Rotation: This rotation was popular in Andhra Pradesh, Jammu and Kashmir, Maharashtra, Punjab and Uttar Pradesh. The mean yield of these demonstrations was 95.17 q/ha of grain and 7.19 q/ ha of pulse.

3. Maize – Wheat- Maize Rotation: The demonstrations were laid out with rotation in Punjab and Rajasthan. The average yield was 79.88 q/ha of grain.

4. Jute–Rice–Wheat Rotation: The average Jute–rice–wheat rotation was followed in West Bengal. The average yield of grain was 65.13 q/ha and that of jute 23.47 q / ha.
5. Rice–Wheat–Fodder Rotation: Rotations with cereal and fodder crops were followed in Haryana and Punjab. The mean yield from such demonstrations was 102.14 q/ha of grain and 259.44 q/ha of fodder.

(C) Demonstration in Rainfed Areas

Demonstrations were also conducted in rainfed areas. The Maize – Wheat rotation in rainfed conditions in Jammu and Kashmir gave an average yield of 44.80 q/ha. In Madurai district (Tamil Nadu), the mixed crop of groundnut and blackgram gave the yield of 14.85 q/ha of pods and 1.90 q/ha of pulse. The demonstrations on cotton in Yeotmal district (Maharashtra) gave a yield of 10.87 q/ha.

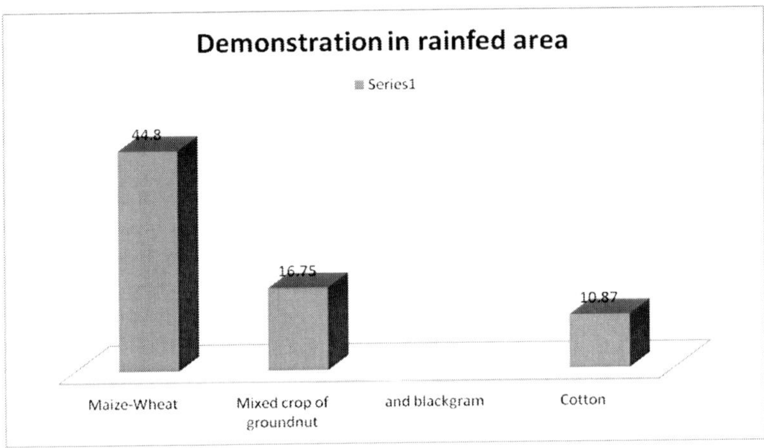

(D) Demonstrations on Problem Soils

Demonstrations conducted on problem soils for demonstrating the technology for reclamation of alkali and acid soils proved successful and gave very encouraging results. With the application of technology available for reclamation of alkali soils, the pH of the soil could be brought down and the cultivators were able to harvest very good yields of the crops which after meeting the cost of reclamation and subsequent cultivation fetched profits ranging from Rs 300 to Rs

500 per hectare. The crops grown on these soils were paddy, sorghum, barley, cowpea and wheat.

Section 2
IMPACT OF NATIONAL DEMONSTRATION ON KNOWLEDGE AND ADOPTION OF IMPROVED AGRICULTURE PRACTICES AMONG FARMERS

Singh and Singh (1974) found that the farmers of villages in which national demonstration was conducted possessed better knowledge (knowledge 39%) than the that of the control village (knowledge 34%) of high yielding varieties of wheat. The knowledge of both the villages was below 40 per cent so the knowledge of the farmers regarding high yielding varieties of wheat in both villages was not satisfactory. This shows that demonstration was not used as educational tool for imparting knowledge of agricultural practices to the farmers. It has to be realized that raising a good crop in the demonstration plot was only a means to educate farmers about the operational details of the practices demonstrated and not an end itself. Greater efforts, therefore, are called for from those conducting national demonstration.

The farmers of demonstration village (24%) had more favorable attitude toward high yielding varieties of wheat than those of the control village (8%). It appears that the National Demonstration Programme did not achieve its objective of changing attitude of the majority of the farmers who had opportunity of seeing it in their own situation as 76 per cent of the demonstration village did not have positive attitude towards wheat varieties. They felt that wheat varieties required very high initial input including capital and too much of care for protection against disease and pests.

The mean percentage adoption in the demonstration village and the control village were 60 and 49 respectively. There was significant difference between the adoption behaviors of the two villages.

Supe and Saloda (1975) measured knowledge and adoption level of farmers of sorghum crop in Akola district of Maharashtra. The respondents were 160 farmers who visited national demonstration plots. The study revealed that national demonstration was effective in increasing the knowledge level of educated, scientific oriented and high socio-economic status farmer participants. Other than the scientific oriented farmers the National Demonstration were not effective in helping the farmers in adoption of the recommended farm practices. This suggests the necessity of supporting the National demonstrations with other extension methods in order to increase the adoption of the farmers.

Pathak et *al.* (1979) carried a field investigation in the districts of 24- Parganas and Hoogly in West Bengal in the year 1976. The respondents were 140 farmers consisting of 70 national demonstration farmers (ND farmers) and 70 non-demonstration farmers (Non-ND farmers). The study revealed that there was significant difference in the levels of knowledge of ND farmers and Non- ND farmers in relation to jute and wheat crops. However, the difference in knowledge was not significant in case of paddy. The difference in adoption level was highly significant between ND farmers and Non - ND farmers in relation to the improved practices of jute, paddy and wheat.

There was significant difference between the attitudes of ND farmers and Non-ND farmers towards multiple cropping. The reaction of the farmers towards multiple – cropping were: (1) the managements was difficult, (2) it was not suitable to farmers of small holdings, (3) it involved greater risk, high initial cost and complicated techniques, (4) the profit in relation was not satisfactory (5) labour was acute during operation time, (6) good farm planning was essential, (7) it demanded continuous work round the year, and (8) It causes inconvenience in attending to social obligations and requirements.

Hirevenkanagoudar *et al.* (1984) studied the impact of National Demonstration on participants and non- participant farmers in 20 villages of five community blocks of Allahabad district in Uttar Pradesh. The total participants were 140 participants and 100 non-participants. The participants farmers had significantly higher knowledge of improved practices of paddy and wheat than the non – participants farmers. Also there was substantial difference between participant farmers than the non – participants farmers with regard to their level of adoption of improved agricultural practices.

In a study by Gupta and Laskar (1988) in three villages in Tripura it was found that 86.66 per cent of the farmers had adopted high yielding varieties after demonstration whereas only 23 per cent farmers were using high-yielding varieties before the National

Demonstration, 33 per cent farmers started to use chemical fertilizer after the introduction of the National Demonstration while only 20 per cent farmers were using chemical fertilizer before the National Demonstration, and 77 per cent farmers used plant protection chemicals after the National Demonstration as against only 10 per cent using before the National Demonstration.

Gaurh and Pathak (1988) found a significant increase in the use of high yielding varieties and fertilizer farmers. They reported the area under high yielding variety (37 to 71%), nitrogenous fertilizer (47 to 70%), phosphate fertilizer (37 to 77%), potassic fertilizer (23 to 58%) and compost (87 to 92%) for paddy cultivation in plots of 24 demonstration farmers In Raipur district increased considerably as compared to the area covered before the programme.

Section 3
IMPLEMENTATION OF NATIONAL DEMONSTRATION

National Commission on Agriculture (1976) in its eleventh report reported the following weaknesses in implementation of National Demonstration. Most of the State Departments of Agriculture, Agricultural Universities and Central Institutes were expected to take the help of the scientists who were specialists in the field of agronomy, soil science, plant protection and water management. However, instead of experienced scientists taking up the National Demonstration work, comparatively inexperienced and junior scientists were made in charge of demonstration. As a result, the original objective of encouraging research scientists themselves to conduct demonstration was not fully achieved. The visit by the subject-matter specialists to the demonstration plots has also taken in as only a routine exercise. Visits by specialists which form an essential aspect of National Demonstrations should have been more frequent so that the expert advice by the specialists posted at operational level was made readily available to the farmers when they faced specific problems. Farmers, however, were deprived of superior expertise because in actual practice, only junior scientists were deputed to participate in these demonstrations. This defeated the very purpose of bringing scientists of proven ability close to the farms, farmers and their problems.

Sufficient attention was not given to pulses, fiber and oilseed crops. Other commercial crops were also not taken in rotation on any large scale. Further vegetable and forage crops did not find a place to the desired extent.

Lack of adequate facilities for soil analysis was another problem. The extension workers failed to get the soil samples analyzed which could enable them to make proper recommendation of fertilizer doses.

Demonstration will have practical value and influence only to the extent people see, hear and read about the results. Even the most modern demonstrations by themselves will benefit only a few people if the results of demonstrations are not adequately shown and understood by the farmers.

The present procedure of conducting demonstrations by providing free inputs does not help effectively in reaching the objective of National Demonstration Programme. Further, when free supply of inputs is tagged on the demonstration work, less emphasis is given to motivate farmers by educating them regarding the advantages of new practice. As a result, the demonstration programmes are not taken seriously in many cases and are treated as routine exercises to be performed. As a result, the objectives of the programme were not fulfilled and achieving the desired objective of motivating farmers through the educational process by using the demonstrations was not achieved.

National Commission on Agriculture (1976) made important suggestions for improving effectiveness of National Demonstration Programme as below after analyzing its operations in the field.

Each research station of Agriculture Universities and ICAR should adopt villages for National Demonstration near to its campus. The demonstration should also include the following subjects: (1) commercial crops, (2) agricultural practices optimizing the use of resources – irrigation, chemical fertilizer and plant protection, (3) Improved practices of animal husbandry, horticulture, fodder crops and forestry. Whole village should be adopted for poultry development and introduction of improved breeding of milch animal since community involvement is essential for such activity. There should be more emphasis of education farmers about scientific method agriculture rather free supply of agriculture inputs to farmers. Literature on scientific agriculture should be distributed in villages. Field days should be organized at demonstration plots. The results of demonstration plots should be widely spread through mass media, print media, extension agencies and group meetings in villages. There should be evaluation of the National demonstration.

Vice-chairman of the National Commission on Agriculture Shri B. Sivaraman said at All India Workshop on National Demonstration in April, 1971 – Development of scientific agriculture was dependent on the growth of two-way traffic between the scientist and the farmers. The National Demonstration provided an opportunity to the scientists to prove on farmers' field the potential of scientific methods in increasing yields. It also enabled the scientists to learn at first hand the farmers' problems in cultivation. The farmers were obviously, to be benefited vastly by the successful implementation and the actual demonstration of the practices their field itself. The programme gave them an opportunity to discuss their problems and get solutions from them. However, this two-way traffic expected in National Demonstration had not really been achieved.

He said that there was a need to give importance to the right choice of crop for crop rotation programmes. He further said that if the scientists involved in national demonstration came across or learnt of some reason for non-acceptance of a particular variety of crop by farmers, this should be passed on the coordinator of the scheme so that the more ready acceptance could be obtained.

J.S. Kanwar, Former Deputy Director General, ICAR, New Delhi (1971) said "The National Demonstration programme was started at a time when the foremost question before the country was providing more food grains. Now the new questions are that besides continuing with onward thrust of more food grain production, we should be able to produce more proteins (pulses), oilseeds (oil and fats), forages and fibre crops. Thus the national demonstration has to include diversified cropping along with one most important food crop as the pivot for the rotation. Secondly, because of the poor economics of coarse grain crops, it is necessary to demonstrate more remunerative crops. Third in view of the high prices of fertilizers, it is essential that most efficient use of fertilizers should be ensured. Fourthly, a watch has to be kept on the changes in pest and disease complex under different cropping patterns so as to take timely action against them. Last of all, an economic analysis of production needs to be made so as to determine the enterprises which are more paying for the farmers. It hardly needs to be emphasized that the full benefit from national demonstration programme will only result if the process of training of farmer and this transference of new knowledge to largest number of farmers make rapid progress and research continuously advances the frontiers of knowledge and improves technology for better and more efficient crop production.

Balar and Patel (1973) studied the working of National Demonstration Programme in 18 districts of Gujarat state during the year 1968-69. The respondents included 34 organizers, 40 demonstration farmers and 40 farmers who visited the demonstrations plot. The study revealed that selection of demonstration farmers and demonstration plots was quite satisfactory. The recommended agriculture inputs were supplied in time in most of the plots and soil analysis was done in all cases. However, some of the organizers were unaware of the objectives and procedure in organizing National Demonstration. There was, in general, lack of coordination between demonstration organizers, extension agencies, local leaders and institutions. It is essential that all farmers of the surrounding villages and all extension officers of the area should know about the demonstration farmers and the sites demonstration plots. Unless they knew well the location of the demonstration plot they would hardly be expected to be present at the demonstration. The basic idea of organizing demonstration is to show to the farmers and extension workers the worth of the demonstrated practice in a particular situation. Displaying signboards at demonstration is an important item as it draws attention of the visitors and passersby. Signboards were prepared by fifteen demonstration organizers. However, there were wide variations in preparing signboards. Some boards were very attractive, but some boards were very small board and not easily visible, while some were signboard on simple iron sheets. Only four organizers had arranged farmers' day on demonstration plots. No proper publicity was given hence the number of farmers and extension officers on the occasion of the farmers' day were very low. Only 63 extension workers were present at farmers' day out of about 2300 extension workers in the state. Twenty six out of 34 organisers did not take any action for publicity of the results of demonstration. The results of the demonstrations should be widely published through radio talks, group meetings, group discussions with farmers and extension workers and photographs etc. None of the organizers could relay demonstration plot results through radio nor took any photograph of crops raised about demonstration. Thus objectives of farmers' education and learning new skills still remained to be achieved. Target yield of crop of nine tonnes per hectare could not be achieved in 80 per cent of the plots. The target yield of nine tonnes per hectare could not be achieved in majority of the demonstrations plot. However, in case of six per cent plots the yield was more than 11 tonnes or more per hectare in the year. This indicated that if demonstrations were properly arranged and all necessary care was taken, it was possible to get the yield of nine tonnes or more per hectare. The organizers generally faced the problems of inadequate transport, lack of timely supply of fund, lack of time and lack of staff.

The authors made the following suggestions in order to improve the programme of national demonstration: (1) Training to demonstration organizer on extension methods, (2) Conducting meeting of demonstration organiser before *kharif* and *rabi* season for discussion about the programme and allotment of demonstration, (3) Preparation of month wise calendar of activities to be undertaken of national demonstration plots by demonstration organizer, (4) Frequent visit of demonstration organizer to demonstration organiser, (5) Timely supply of agriculture inputs to the farmers, (6) Wide publicity of results of demonstration, (6) Sign board at demonstration plots, (7) Transport facility to demonstration organizer, and (8) Coordination between research institutes and extension agency

The study of Mishra and Singh (1970) in Bihar reported that 35 out of 40 demonstration officers had a definite plan of conducting the demonstration. Although demonstrations were conducted by teaching and research staff, extension workers could not be done away with. The involvement of extension workers is a must for the programmed to achieve the successful results. Twenty six out of 40 demonstrating officers did not send any information to the Block Personnel regarding the demonstration they conducted. Most of the demonstrating officers gave thought to ascertain desirable attributes in the selection of a demonstrating farmer. However, several of them ignored a few important points in selection of demonstrating farmers. Only 24 out of 40 demonstrating officers considered that the demonstrating farmer should maintain a proper record of his demonstration. In most cases the plots for demonstration were chosen carefully.

However, nearly one fourth of them did not ensure that the plot selected was be near the village, that it represented the characteristics representation of the area, that the plot was owned by demonstrating farmer himself, and it was well protected from stray animals and was cultivated by the demonstrating farmer himself. These are actually very important considerations in the successful execution of demonstration. The demonstrating officers might have ignored these points because of lack of knowledge about how a successful demonstration is to be conducted rather than carelessness. Soil analysis which is an essential precondition, was done in case of only 11 out of 40 demonstrations and the dose of fertilizers in the remaining 29 demonstrations was decided on the basis of general recommendations rather than a scientifically obtained measure. Thirty two out of 40 demonstrating officers displayed signboards at the site of their demonstrations. The presence of block personnel and farmers at the time of harvesting, threshing and weighing of produce of demonstration plots is very important. The ultimate purpose of conducting demonstrations is to convince the farmers of the advantages of improved agriculture practices demonstrated and unless the farmers themselves see the result of these practices they will not accept them. The percentage of demonstrating officers who reported the presence of block personnel and farmers at harvesting and threshing stages varied between 45 and 55 and 55 and 95 respectively. Ten out of 40 demonstrating officers did not report the presence of block personnel and farmers at any stage of these demonstrations. Twenty out of 40 demonstrating officers had no problems with regard to arrangements and the material required for demonstration. The remaining 18 demonstrating officers experienced difficulties *viz.* inadequate and untimely allotment of funds and unavailability of necessary plant protection chemicals and seed in time. Out of 40 demonstrating officers 31 tried to know the reaction of demonstrating farmers. The farmers at whose farms demonstrations were carried out had the following views about the demonstrations: (1) the practices demonstrated needed much more and supervision, (2) the cost involved was high, (3) profit in relation to the cost was not sufficient and (4) the practices were complicated and risky.

The authors recommended the following measure for improvement of National Demonstration: (1) Adequate funds and advance supply of agriculture inputs, (2) Collaboration with extension agency, (3) Wide publicity and follow, (4) Whole time officers with provision of vehicle, (5) Regular supervision by higher authorities, (6) Advance communication of various recommendations (7) Conducting of demonstration in each community development block and (8) Advance preparation of crop schedule.

Sharma and Jha (1972) studied the perception of national demonstration by specialists (20), extension personnel (20) and demonstration farmers (20) in Rajasthan. All the respondents were aware of the implementation of national demonstration in Rajasthan. Thirty five per cent of the specialists stated that national demonstration were laid on uniform pattern at the National level while only 25 per cent of the extension personnel mentioned it and none of the demonstrating farmers. Only about 15 per cent of the specialists were aware that conducting National Demonstration was their responsibility. Against this more of the extension personnel (55%) were aware of this. Even among the demonstrating farmers there was better awareness (45%) about this. This reveals that most of the specialists did not perceive their central role in conducting national demonstration correctly. Only 40 per cent of the specialists were aware of the package of practices demonstrated in national demonstration. This again leads to the inference that the specialists did not have clear understanding of the programme.

The idea of minimum yield targets was known to 35 per cent of the specialists, 55 per cent of the extension personnel and 70 per cent of the demonstrating farmers. Fifteen per cent of the specialists, 50 per cent of the extension personnel and 60 per cent of the demonstrating farmers reported that in national demonstration the stress was mainly on training the farmers about the package of practices.

There was a general agreement among about half of the respondents groups on four steps required for the demonstration namely, (1) selection of farmer, (2) selection of sites, (3) analysis of soil sample, and (4) carrying out the recommended package of practices. This shows out that the steps emphasized and commonly understood by 30 to 60 per cent of the specialists pertained to the operational part only and that too only for pre-demonstration and execution stages; and not for the follow-up stage. Some other steps basically related to making the farmers understand the purpose of national demonstration, teaching during demonstration on field days, farmers' visit to the site and effective follow-up were not considered to be important by most groups of the respondents. Only a few of the prescribed steps were mentioned by the specialists. This might be because planning and targets for these demonstrations were decided by higher authorities and so specialists had not much say in making decisions. All of them reported that they carried out the demonstrations as per instructions received from their superiors.

Only 25 per cent of the specialists and 40 per cent of the extension personnel received guidance from their superiors. In no case, training on different aspects of demonstration programme was given to the specialists and extension personnel. Neglect on the part of higher authorities with respect to providing training or instructions might be the reason of having poor perception of concept, objectives and process of national demonstration by the persons who conduct such demonstration.

It was found that farmers' interest was less in learning technical details of the practices demonstrated and more in acquiring free seeds and fertilizer.

The study brought to the fore that (1) personal characteristics of the farmers like their interest and enthusiasm, willingness to cooperate, their status as a progressive farmer commanding respect of the villagers; (2) having enough farm resources and irrigation facilities at a good site; and (3) their personal terms and obligation were the main criteria followed for selection of the demonstration farmers. While selecting the demonstration farmers on the basis of desirable personal characteristics, a suitable site with irrigation facility is justified, there was no justification to select farmers because of personal preferences.

The main criteria considered by the specialists for selecting sites were: (1) irrigation facility, (2) nearness to the village, and (3) easy accessibility. However, it was reported by some specialists and extension workers that the land offered for demonstration was fallow. Only 25 per cent of specialist, 65 per cent of the extension personnel and 55 per cent of demonstration farmers mentioned about the field days. This shows that National Demonstration Programme as an educational tool was not fully appreciated by the specialists.

70 and 75 per cent of the specialists and the extension workers respectively stated that farmers were satisfied with national

demonstration. Only 60 per cent of demonstration farmers themselves were satisfied.

Only 35 per cent of the specialist stated that the results were given publicity whereas 85 per cent of them stated that the results were given publicity through field days and 55 per cent of them reported that the use of other source like radio and printed material for the sake of publicity.

Among the difficulties mentioned by the specialists important ones were: having little time at their disposal to visit the demonstration plots for training the farmers and delayed supply of inputs for conducting the demonstration.

The study suggested that training should be imparted to the specialists who are to plan and conduct demonstrations. This will help in creating a clear and uniform perception in them about the concept, the objectives and the process of national demonstration. An understanding has to be developed among the specialists that conducting national demonstration is not an end in itself but means to achieve the end of educating the farmers about the latest technology in agriculture. Careful systematic planning is necessary. Adequate follow-up measures, wide publicity and mass educational efforts on the part of organizers of such demonstrations are not less important. In no case should these demonstrations be conducted by the operational staff like village level workers, with specialists only supervising the operation. Instead, each and every step of a national demonstration should be performed by the specialists themselves in the presence of as large a number of farmers as possible. A nationwide study should be undertaken to assess the working of national demonstration programme.

Behera and Sahoo (1975) found that only three-fourth of sample farmers in eight villages of Puri district knew about national demonstration. Even less than half of the farmers visited the plots although quite a good number of them had seen the sign board. They recommended the following measures for enhancing the effectiveness of National Demonstration. The demonstration farmer and site should be selected by the specialists in consultation with the local block staff and the village farmers. Prior information should be circulated three to seven days before the field day and other meetings of National Demonstration. The subject matter specialists should pay more attention to educational objectives of the programme than to achieving the yield target. They have to encourage other farmers from the same village as well as the neighboring villages to participate more in the programme.

REFERENCES

Anantharayana, R., Eswarappa, H. and Patel, V. N., "National Demonstration show the way to Increase Crop Yields in Chitradurga District, Indian Farming ", XL, 6, pp. 15-16, 1990.

Annonymous, Role of National Demonstration, Indian Farming, XXI,3, pp.46, 1971.

Balar, N.H. and Patel, I.C., "Evaluation of the National Demonstration Programme in Gujarat State", Indian Journal of Extension Education, IX, 1 & 2, pp. 27-36, 1973.

Behra, C. and Sahoo, M.S., "Impact of National Demonstration on Adoption of Agricultural Practices", Indian Journal of Extension Education, XI, 17 2, pp. 32-35, 1975.

Dadwal, K.S., Dutt, B.K., Dadhwal., Gupta, S.K. and Lascar, S., "Yield Gaps Between Demonstration and Farmers fields in Tripura", Indian Farming, XXXIII, 1, pp.13, 1983.

Dutt, B.K., Dadhwal, K.S. and Gupta, S.K., "National Demonstrations in Tripura", Indian Farming, XXXIII, 9, pp.45-47, 1983.

Gaurh, R.S. and Pathak, P.K., "Impact Analysis of National Demonstration in Poplarisation of Rice Technology on the Farm of Demonstrating Farmers", Maharashtra Journal of Extension Education, IV, pp. 155-157, 1985.

Goydani, B.M., Awasthi, A.K., Ghode, B.D. and Sawarkar, S.D., "National Demonstrations in Madhya Pradesh Show the Way to Progress", Indian Farming, XXXVII, 9, pp.31-40, 1988.

Gupta, S.K. and Laskar, S., "Impact of National Demonstration", Indian Farming, XXXVI, 10, pp. 23-25, 1988.

Hindu, 5, 283, pp. 15, November, 2015

Hirevenkanagoudar, L.V., Chitambar, J.B. and Chaubey, B.K., "Impact of National Demonstration on Participants and Non-participants", Indian Journal of Extension Education XX, 1 & 2, pp. 76-78. 1994.

Kanwar, J.S., "National Demonstration: Retrospect and Prospect", Indian Farming, XXI, 6, pp. 13-17, 1971.

Mahapatra, I.C., "National Demonstartion Trials", Indian Farming", XIX (3), pp.24-27, 1969.

Malhotra, O.P. and Katyal, S.K., "National Demonstration in Haryana Show the Way to Progress, Indian Farming", XXXVI, 9, pp. 35-36, 1986.

Mishra, A.P. and Singh, R.P., "An Analysis of National Demonstration Programme", Indian Journal of Extension Education, VI, 3 and 4, pp. 63-70, 1970.

Pathak, S., Pal, M.K. and Guha Roy, M.K., "Impact of National Demonstration on Knowledge, Attitude and Adoption Level of Farmers in West Bengal", Indian Journal of Extension Education, XV, 1 & 2, pp.49-53, 1979.

Report of National Commission on Agriculture, Part XI Research, Education and Extension, Government of India, Ministry of Agriculture and irrigation, New Delhi, pp.239-246, 1976.

Shanmugasundaram, A., Diwakaran, K., Natrarajan, M. and Shanmugam. "National Demonstration Impresses Coimbatore Ragi Growers", Indian Farming, XVI, 4, pp.7-8, 1966.

Sharma, K.D. and Jha, P.N., "National Demonstration as Perceived by Specialists, Extension Personnel and Demonstrating Farmers", Indian Journal of Extension Education, VIII, 3 & 4, pp.9-17, 1972.

Sharma, S.K., Nayar, B.B. and Gupta, M.P., "National Demonstrations Show the Way", Indian Farming, XXXII, 7, pp.17-19, 1982.

Singh, K.M.P. and Singh, R.P., "Impact of National Demonstration in Adoption of High Yielding Varieties of Wheat", Indian Journal of Extension Education, X, 1 & 2, pp. 65-67, 1974.

Singh, M.P., Yadav, R.N., Singh, G.R. and Singh, R.R., "Extending Improved Crop Production Technology through National Demonstration Project", Indian Farming, 44, 2, pp. 29-32, 1994.

Subbiah, K.S., Ramanatha, O.S. Anavara and Pillai, L., "National Demonstration of Ragi and Bajara Impresses Farmers", Indian Farming, XIX, 6, pp. 29-31, 1969.

Supe, S.V. and Salode, M.S., "Impact of National Demonstration on Knowledge and Adoption Level of Farmers Participants", Indian Journal of Extension Education, XI, 1 & 2, pp. 37-39, 1975.

Swaminathan, M.S., "National Demonstration", Indian Farming, XVI, 6, pp. 68-70, 1966.

Swaminathan, M.S., "The Purpose and Philosophy of National Demonstration", Indian Farming, XXI, 6, pp. 11-12, 1971.

■■■■

CHAPTER - 4

GUIDELINE OF LAB TO LAND PROGRAMME

The **Lab to Land Programme (LLP)** started in 1979 -80 was one of the important programme of the **Indian Council of Agricultural Research (ICAR), New Delhi.** In this chapter, we will take a look at the working of the programme. The guideline for Lab to Land programme dealt with by **Saxena *et al.*** (1979) in the paper "the *ICAR Golden Jubilee Lab to Land Programme for Transfer of Technology"* is presented below.

ICAR, New Delhi which was established on 16th July 1929, completed 50 years of service that (1979) year. To commemorate the completion of 50 years of service by the ICAR, the then President of the ICAR Registered Society, Shri Surjit Singh Barnala, then Minister for Agriculture and Irrigation, constituted a National Organising Committee for ICAR Golden Jubilee Celebration under the Chair-manship of Shri Bhanu Pratap Singh, then Minister for Food and Rural Development and Vice-President of ICAR. On this occasion the President suggested that it would be appropriate if during the Golden Jubilee Year, special emphasis was placed on the transfer of technology which could have desirable impact on the development of agriculture in the country and economic well-being of the farming community.

The National Organizing Committee for Golden Jubilee of the ICAR decided that the most important activity during the Golden Jubilee Year (1979-80) would be massive effort for the transfer of technology from laboratory to farmer's fields. This was named as Lab to Land Programme.

The programme planned to select 50,000 farm families belonging to the small and marginal farmers and landless workers and aimed to assist them in developing and implementing individual farm plans for improving the entire farming activity. The idea was also to try to release technologies which could help in diversification of farm labour and introduction of supplementary sources of income such as field of animal husbandry, sericulture, apiculture, etc.

It was visualized that experimental programme would enable the participating Institutes/Agricultural Universities to accelerate and intensify their existing efforts made to transfer technology without creation of any additional infrastructure. This could mark their contribution to the commemoration of Golden Jubilee of ICAR, New Delhi.

Another very important objective of this experimental programme was to develop a strong feed-back mechanism which could enable the scientists to understand problems and needs of the farmer and to identify the constraints in the adoption of new programmes of technologies transfer to the farmers.

Hyderabad Workshop

As a follow-up action, a workshop on ICAR Golden Jubilee Transfer of Technology Programme for Rural Development was organized at Hyderabad from 3-5 January, 1979. The workshop made following recommendations as discussed below.

1. Selection of centers for transfer of technology; criteria to be used and mechanism of operation

It was recommended that the Institution/Agency involved in this programme should ensure that the number of families adopted by them is effective. In selecting the centre, preference may be given to the Institution/Agency already involved in similar agricultural development work.

It was agreed that in this programme, there would be full involvement of the ICAR Institutes and its various regional stations, sub-stations/centres/KVKs/(AICRPs)/(NDCs)/(ORPs). In addition, some of the selected voluntary organizations and affiliated Agricultural Colleges would also be involved.

There were about 300 such centres under ICAR Institutions and about 460 centres under the Agricultural Universities, in addition to about 1000 affiliated agricultural colleges including voluntary agencies*. It was been estimated that ICAR centers would handle about 20,000 farm families and the Agricultural Universities centers would handle an equal.

Number and the balance 10,000 were to be handled by the voluntary organizations and the affiliated agricultural colleges. The above estimates were on a tentative basis and the actual selection of centres and allocation of farm families would take into consideration various amenities, services etc. available.

Agricultural Universities would be the co-ordinating agency for selection of farm families in respect of its own centers, and other AICRPs, ORPs, etc under their administrative control. Similarly, ICAR Institutes would select farm families in respect of various centers under their control and co-ordinate their programme.

In selecting the families, assistance available from Agricultural Universities, ICAR Institutes, State Department of Agriculture and other agencies, including voluntary agencies, would be taken into consideration.

Likewise voluntary agencies, while selecting the farm families the farm families would associate the local Agricultural Agencies/ Institutes and State Department of Agriculture, etc. and select the number of families depending upon their available resources.

Necessary technical expertise and support would be made available to the voluntary agencies by the concerned ICAR Institutes/ Universities.

2. Selection of participating farm families

It was recommended that selection of Block/Villages/Families should preferably be from intensive agricultural development blocks.

---

* Krishi Vigyan Kendra (KVK), All India Coordinated Research Project (AICRPS), National Demonstration Centre (NDC), Operational Research Project (ORP).

While selecting villages, preference would be given to the villages having large proportion of families belonging to the category of small farmers, marginal farmers, landless laborers, scheduled tribes and scheduled castes.

The approach during transfer of technology should be on a cluster-village basis. Proximity to the implementing agency may be kept in mind for effective supervision/implementation of the programme. In selecting farm families following broad criteria should be kept in mind. Preference should be given to farm families owning less than hectares of irrigated land. While selecting village's existence of a school, a co-operative society, an effective village panchayat, etc. should be kept in mind.

In this programme, farmers' families should be taken as a unit and related programmes for farm youth, farm women and children should be initiated simultaneously. Creation of suitable marketing infrastructure and facilities well in advance to match the anticipated increased production possibilities under this programme was also strongly recommended.

Lab to Land Operational Manual

The **Lab to Land operational manual** containing brief information on all technologies available would be supplied to all participating centres in this programme. It was suggested that this manual should highlight information in respect of proven and viable technologies available for small and marginal farmers.

For every item included in the Lab to Land manual, the concerned Institute/Agricultural Universities would develop and provide suitable extension literature in simple language for use by village level workers and participating farm families. The Institute/ agricultural University would provide on the request the extension literature, to the participating centres.

It was decided to suitably modify and simplify the performance for benchmark survey already developed which would be provided to the participating centre as early as possible.

For continuous monitoring and evaluation of the programme, a simple, and computer–tailored pro forma may also be evolved. Based on the assessment/evaluation of the projects, suitable reports should be published and circulated for effecting necessary mid –course corrections.

Organisation of National Extension programme

An intensive Extension Fortnight could be organized preferably during the first fortnight of September, 1979. It was agreed that Universities and Institutes would choose suitable dates on local condition and convenience.

It was agreed that the following programme may be undertaken during the ICAR Golden Jubilee Year.

- Organisation of Agricultural Exhibitions at National and State Institutes and University levels.
- Organisation of Field Days at various sub- stations/ Campuses of ICAR Institutes and Agricultural Universities.
- Organisation of Mobile Exhibitions by Agricultural Universities and ICAR Institutes.
- Mass contact programme by scientist and students of Agricultural Universities and ICAR Institutes.
- Organisation of lectures by scientists in rural schools duly supported by audio-visual aids, film shows, etc. to highlight the potential and possibilities of adopting agricultural technology.
- During this fortnight broadcast and telecast of special programmes/features with a caption under the programme 'Lab to Land may be arranged on the Radio and Television.
- In addition to extension literature to be provided and distributed by the various agencies during this fortnight, efforts should be made to bring out special supplements/ regular columns in farm journals and newspapers and other local publications.
- Inviting visiting farmers/Professors to visit agricultural Universities/ICAR Institutes to discuss with scientists their successful farming experiences.
- Organisation of debates at institutional, State and National levels.

- Organisation of National Seminar for policy –makers for improving organizational efficiency and extension services
- In order to have a feed-back, a reverse Land to Lab Programme should be organized for identifying constraints in the adoption of farm technologies.
- Earth Satellite Station in Delhi producing T.V. programmes for the centres like Udaipur, Muzaffarpur, Kanpur, Lucknow, etc., should keep in mind the timeliness and applicability of the recommendations for these areas.

While celebrating Extension programme, every programme may be identified by displaying a common banner and a design which may be provided by ICAR in both English and Hindi language. The banner in the regional languages should be prepared by the concerned implementing agency.

It was recommended that the entire research/teaching and extension staff of the ICAR Institute/University may be involved in this programme.

It was recommended that the necessary funds and other inputs for implementation of activities like holding of exhibitions, provision of literature, debates, mobile exhibition units, etc. should be made available to the various implementing agencies.

Organization of special programmes for women and children should also be undertaken during the ICAR Golden Jubilee Year which would synchrony with the International Year of the Child.

General

ICAR directed various organizations like Agricultural Universities/ Institutions/voluntary agencies, etc to earmark separate groups of scientists /technologists for implementation of the programme with selected farm families. These institutes and agencies helps in the organization of exhibitions and under-taking training of farm youths. These scientists were expected to devote minimum 1/3 of their total time for Lab to Land Programme.

Necessary facilities in terms of transport, contingencies, etc. were to be made available well in time so that training programmes, identification of families is done before the kharif in 1979.

It was recommended that suitable correspondence courses should be organised to farmers on special subjects on payment of National fees at a part of this Programme.

For effective implementation of the entire transfer of technology programme, the ICAR was also to evolve a suitable financing pattern for any special items under this programme.

During the ICAR Golden Jubilee Year avenues for setting up of Broadcasting and TV Station at some of the selected Agricultural Universities/ICAR Institutions for development and transmission of special programmes was to be fully explored to ensure the success of various Agricultural Development Programmes in the country.

General Guidelines

On the basis of the recommendations made in the ICAR Workshop/ held at Hyderabad in January, 1979, a guideline was prepared which was issued to all the Institutions/agricultural Universities and other agencies implementing the Lab to Land Programme. The following guidelines were issued as discussed below.

Each Agricultural University, Regional Research Station, ICAR Institute, and voluntary agency would constitute a team of specialists comprising of scientists drawn from different disciplines of agriculture, Veterinary Science, Fisheries and Home science, etc.

The multi-disciplinary team would be headed by a full- time co-ordinator provided by the implementing institution/agency.

Each centre was to adopt 50-100 families dispersed in backward villages covered by SFDA, MFAL, DPAP, etc. programmes. These villages would be selected in consultation with the Project/ Agricultural Officers/BDOs and would be as far as possible, in the vicinity of Regional Centre/College/ICAR Institute. Farm families selected under the programme should own 1-2 hectares of land; should be willing to participate in the programme; willing to provide necessary information and data as required; willing to undertake training for implementation of the programme and should have a clear record in the village and it's social life.

A comprehensive survey about the farming systems, level of farming, material resources, socio-economic conditions etc was be carried out.

- The blueprint for the development of adopted would be prepared with an emphasis on improvement in the farming systems, employment potential and living conditions.
- Each specialist was to be entrusted 10 families and would be held responsible for procurement of grants, loans for various innovations required in the blue print for family development.
- The Co-coordinator/Team Leader was expected to make maximum use of grants/loans provided by the Rural Development Departments through Small farmer SFDAs, MFALs, DPA, IBRD, etc., programmes for undertaking community infrastructures and organizations of managerial and other training programmes.
- The members of the implementation team were to maintain regular and frequent contacts with the adopted families which should be at least once in a week for providing solutions to the problems in implementation of the rural development programme.
- Special programs for women and children were to include so as involving all members of the elected farm families in transfer of technology.
- Vocational training programmes for improving income generating skills, such as basket weaving, poultry keeping, animal husbandry and other skill like proper utilization of food and fodder by the farming families were to be initiated for landless laborers.
- As regards the funding of the Lab to Land programmes, it had been felt that the subsidy for raising crops on the farms of selected farm families should be in line with the pattern followed under National Demonstrations project. The subsidy thus available to implement the crop production programme for these farm families was primarily to be used for supply of critical inputs–inputs which normally a farm family could not afford, i.e. additional dose of fertilizer, insecticides, pesticides, improved seeds of varieties etc. and should be provided in kind only.

- These funds would only be made available to the operational families on the recommendation of implementing agencies such as agricultural universities and ICAR Institutes. The selection of villages and families was to be confined to the blocks covered under the Integrated Rural Development Programme being implemented by the Department of Rural development in various states.

For identification of beneficiaries under Lab to Land Programme of the council, the following definition of small and marginal farmers and landless laborers was used:

1. **Small farmers and cultivators** with land holding below 5 acres. In case of irrigated land, the ceiling was be 2.50 acres.
2. **Marginal farmers – cultivators** with land holding up to 2.50 acres – incase of irrigated land, the ceiling was be 2.25 acres.
3. **Agricultural laborers – cultivator** without any a home-stead and drawing more than 50 per cent of their income as agricultural wages.

In case of beneficiaries under Drought Prone Area Programme (DPAP), the definition given by the Department of Rural Development is to be followed since the land limits under this programme differ from state to state.

As indicated earlier, for crop demonstrations, the pattern of providing subsidy should be that of National Demonstration Project. In all the financial schemes meant for individual families of rural community, the guidelines provided by the Department of Rural development for Integrated Rural Development Programme (IRDP) were to be followed strictly.

The number of crop demonstration to be laid out by the implementing agencies under the Lab to Land programme were related to the number of families selected under this programme. In principle, there should be one demonstration per each selected family. Farm family plans for each of the families covered under the programme were to be prepared and implemented by the implementing agencies.

Programmes for agricultural labourers, scheduled castes and scheduled tribes were the same as envisaged by the Department of Rural Development, Government of India.

The council agreed to provide grant for this programme according to the pattern of assistance approved under the ICAR Golden Jubilee Lab-Land Programme subject to budgetary limitations imposed by the Government of India from time to time. The pattern of assistance would *inter-alia* include the following:

Items	% amount (Rs)	Rs.
1. 100 crop demonstrations with 2-3 crops per year (on the same land –one acre)		
a. Irrigated command areas (Rs. 200 each for the first and second crops & Rs. 100 for 3rd crop)	30 × 500	15,000
b. Assured rainfall areas (Rs. 200 each for 1st & 2nd crop & Rs. 100 for 3rd crop)	35 × 500	17,500
c. Dryland agriculture area (Rs 200 each for 1st and 2nd crop)	35 × 400	14,000
Total		**46,500**
2. Organisation of field days, exhibitions, extension fortnight		3,000
3. Production of popular literature (Five leaflets)		1,000
4. Travelling allowance		3,000
5. Pol charges		3,000
6.Training of participants farm families and orientation course for technical staff		2,500
Total		**12,000**
Rs. 59,000 per 100 farm families		

1. The demonstrations on problem soils would have a subsidy of Rs 1000 per acre per year as available under National Demonstration Project. However, this amount would be

within the overall limits prescribed over for crop demonstration.

2. The demonstration on fish farming /composite fish culture will also have the subsidy at the rate of Rs.500 per demonstration. The size and number of demonstrations would be determined by the implementing agency depending upon the availability of ponds, etc.

3. The utilization of the above money or substitution from one head to the other is left to the discretion of the implementing agency. An addendum to the above guidelines after due consideration in 2nd Zonal Co-coordinators meeting hold on 20th November, 1979 was issued to all the implementing agencies.

Addendum to the Guidelines of Lab to Land Programme

List of Critical Inputs Eligible for Receiving Assistance under Lab to Land Programme (as approved at the 2nd Zonal Coordinators' Meeting held at new Delhi on 20-11-1979).

List of Critical Inputs Permissible

Items	**Permissible Critical Inputs**
Crops	1. Purchase of seeds/seedlings, saplings
	2. Purchase of fertilizes.
	3. Purchase of insecticides, pesticides
	4. Soil conservation, land shaping
Dairy	1. Health cover to cross-bred animals (Immunization & medicine)
	2. Inputs for raising fodder crops.
	3. Concentrate supplement including mineral mixture.
Piggery	1. Supplementing feeds for piglets.
	2. Health cover (Immunization & medicine)
Poultry	1. Health cover (Vaccination & medicine)
	2. Feeding charges supplement.

Items	Permissible Critical Inputs
	3. Supply of chicks.
Sheep & Goat	1. Health cover (drenching, vaccination, medicine, etc.)
	2. Enrichment of poor quality roughage.
	3. Post –natal care.
	4. Shearing.
Fisheries	1. Fry and fingerlings.
	2. Supplementing feeds.
	3. Mahuva oil cake, chemicals foe eradication of predators
Sericulture keeping	1. Purchase of chandrikas, trays, tray stand, mulberry plant, nurseries for farmers already engaged in this enterprise.
	2. Purchase of bee-hives and wax foundations.

Note:

1. Critical input in kind up to a ceiling of Rs.500 per family. For problems of soil the ceiling was upto Rs.1, 000.

2. Maximum permissible assistance per family would be limited Rs. 500 per agriculture year, irrespective of critical inputs used under various heads.

A Memorandum of Understanding was entered into by the ICAR with participating agencies. A copy of the list of permissible eligible for receiving assistance under Lab to Land Programme along with addendum was sent to all the Transfer Technology Centre (TTCs). In allocating farm families due care were taken to ensure that the distribution of 50,000 farm families is done with the aim of distribution over the entire country, which been divided for this purpose into 8 agro- climatic zones.

For implementation of the Programme, 8 Co-ordination Units had been set up.

The programme inter-alia involved conducting of a Bench Mark Survey of the selected farm families for which a suitable Benchmark Pro forma was developed by a special task force team, copies of which were provided to the Directors along with the first and second drafts of the Lab to Land Manual. The initial Bench Mark Survey was conducted by the concerned TTCs. The task force also developed a quarterly progress report Pro forma for monitoring and evaluation of the programme, copies of which had been sent to all the TTCs for sending the progress report of the first quarter. This would enable to the release of the second installment of money sanctioned for their Technology Transfer Centre's under Lab to Land Programme.

Another important part of the programme was the preparation of individual farm plan for each of the participating farm families by the Technology Transfer Centers, which was under way.

The lab to Land Manual incorporating the available tested and proven technology was brought out university-wise and institute-wise. Copies of manuals were provided to all the Technology Transfer Centers (TTC) in the hope that this would provide a ready reference to all Technology Transfer Centers in the country and keep them informed about the availability of technologies and enable them to get in touch with the concerned source-material institution for utilization in their areas.

The participating institutions agreed to make available on the request suitable extension literature to any other TTCs. This extension literature along with material on package of practices is also available for the participating farm families.

A provision of Rs 3 crores was made for the agricultural year June 1979 to May, 1980 for Lab to Land Programme in respect of 50,000 farm families in accordance with the guidelines, which were amended to include animal husbandry programmes.

In addition, a provision of Rs. 20 lakhs was made for expenditure in connection with the creation of 8 Zonal Coordination Units, which were charged with the responsibility of planning, implementation, monitoring and evaluation of the Lab to Land Programme. Likewise, a provision of Rs 8 lakhs was made for creation of supporting cell at ICAR Headquarter for the entire Golden Jubilee Programme, which includes Lab to Land Programme, Integrated Agricultural information Support and the ICAR International Symposium.

The ICAR Headquarters Golden Jubilee supporting cell was functioning under the overall charge of the Deputy Director General (Soil) who was responsible for overall coordination and collation of the entire ICAR Headquarters Golden Jubilee Programme.

The post of Scientist in Economics was created under this Programme, whose services were placed with the Indian Agricultural Statistics Research Institute for the purpose of monitoring the progress of the Lab to Land Programme in 8 Zones using the computer facilities available with them.

Similarly, the integrated agricultural Information Support Cell at the ICAR Headquarters was coordinate work of publicity and public relations through the use of the existing media/ facilities of Government of India/ states and the Press, etc. for giving the widest publicity to the progress of work under this programme. High priority and importance was given to this part of the programme as it was felt that exchange of information on the successful implementation of this programme relating to improvement in the social and economic lives of the participating farm families would eventually enable the extension of this programme from its present experimental level to the national level, which would have far reaching implications for the success of this idea of Transfer of Technology to the farmers in the field of agricultural development.

■■■■

CHAPTER - 5

Research Studies on Lab to Land Programme

The scientists of agricultural universities and research institutes of Indian Council of Agricultural Research, New Delhi implemented improved agricultural practices at the small and marginal farmers' fields under Lab to Land Programme. They also assessed impact of improved agriculture practices and published the findings in research journal. The research papers deal with how improved agriculture practices were selected for the programme, what extension methods were used to educate the farmers about improved agriculture practices and how far agriculture productivity obtained under the improved agriculture practices was higher than farmers' method of cultivation. The salient findings are discussed below.

Section 1
EXTENSION METHOD OF FARMERS' EDUCATION ABOUT IMPROVED AGRICULTURE PRACTICES AND THE OUTCOME OF IMPROVED AGRICULTURE PRACTICES

The research institutes made a multidisciplinary team of scientists for implementation of the LLP programme in villages. The scientists selected the villages and farmers for the programme. They made a survey to have knowledge of the socio-economic status of the farmers and local farm practices before the programme. On the basis the survey, they decided appropriate and relevant agriculture practices for the programme. They used several extension methods (individual contact, group contact, training, demonstrations, extension literature, field, exhibition and film show) to upgrade the knowledge and skill of the farmers about improved agriculture practices selected for the programme. The agriculture inputs envisaged in the programme were provided to the farmers. In initial stage of the programme the farmers were not enthusiastic about the programme, but after seeing the impact of improved agriculture practices in field they cooperated with the scientists.

There was substantial increase in agriculture productivity of crops due to improved agriculture practices in LLP villages by the research institutes (Table 5.1). There was high yield (q/ha) of crops owing to improved agriculture practices : paddy 91, wheat 53.37, pearl millet 29.28, maize 35, potato 150, tobacco 24.04, cassava 42.80, groundnut 28, finger millet 20 and soybean 15.

The yield of every crop under improved agriculture practices was very high than the farmers' method of crop cultivation. The increase in crop yield improved agriculture practices over farmers' was method of cultivation of crop from 18 to 350 per cent.

The percentage increase in the yield of various crops of improved agriculture practice over farmers' method of crop cultiv-ation was: paddy 18 to 350, wheat 21 to 47, maize 42 to 311, tobacco 35 to 54, pearl millet 23 to 133, potato 81 to 150, groundnut 145, finger millet 222, soybean 50.

The improved agriculture practices of crops-paddy, wheat, pearl millet, maize and potato were implemented by the research institutes in the programme. The percentage increase in the yield of these crops under improved agriculture practices over the farmers' method of cultivation was not same in the LLP villages in the country. This is because the yield of crops depend on the soil type, their nutritious status, rainfall pattern, irrigation facility, quality of irrigation water and other management practices that differ to a great extent in the various parts of the country. The same is true for yield of crops with the farmers' practices because farmers of different area used different quantity of inputs and management differ from farmer to farmer in the same locality.

The cropping intensity was increased in the programme due to the effort of the scientists. Pearl millet for cultivation after tobacco as a single crop in kharif season was introduced in LLP villages by Central Soil and Water Conservation Research and Training Institute's Research Centre at Valsad (Gujarat). Multiple cropping by taking *Rabi* crops like – oat, wheat, mustard, pea, potato after rice and maize grown in kharif season was started by ICAR Research Complex, North Eastern's Centre, Saramsa (Sikkim). Intercrop – groundnut wit cassava- was started in the programme by Central Tuber Crop Research Institute, Trivandrum.

With respect to animal husbandry, rearing of goat, pig and poultry with scientific method in the programme gave one and half times to three fold higher productivity in comparison to local breed. Similarly package of practices of poultry quality birds of variety (White Leg Horn), balance and economy ration, deworming and vaccination introduced in the programme increased egg production from 60 to 158 year – an increase of 163 per cent in production.

Table 5.1. Improved agriculture practices on crop and its impact on crop yield in LLP villages

Research Institute/ Improved technology	Crop	Improved practice (q/ha)	Farmers' method (q/ha)	% incr-ease over farmers' method
Punjab Agricultural University, Ludhiana (Bhullar *et al.*, 1988)				
Wheat: Improved seed (Sonalika), seed rate (88-100 kg/ha as against 175-200 kg/ha), Arelan weedicide	Wheat	49	36	36
Paddy: Improved seed (PR 106), seed rate 20 kg/ha as against 30-37 kg/ha farmers' practice) seed treatment (Aldrin + Streptocycline), zinc sulphate	Paddy	91	48	89
2. **Central Soil and Water Conservation and Training Institute, Dehradun (Tyagi *et al.*, 1989)**				
Non-monitoring technology- Timely sowing of quality seed in row across slope, optimum plant population, timely interculture, proper placement of seed and fertilizer, disposal of excess runoff water from the field, high yielding variety	Upland paddy	19.74	16.41	20

Research Institute/ Improved technology	Crop	Improved practice (q/ha)	Farmers' method (q/ha)	% increase over farmers' method
	Paddy irrigated	24.32	20.63	18
	Wheat	32.39	22.00	47
	Maize	26.56	18.75	42
3.Central Soil and Water Conservation and Training Institute's Research centre at Valsad (Gujarat) (Chandra *et al.*, 1991)				
Introduction of tobacco crop after pearl millet, improved variety (Anand 2 and Anand 1190, timely transplanting, FYm 25 cart load, ammonium sulphate 250 kg and urea 222 kg/ha, 2-4 irrigation from collected water.	Tobacco	19.24 to 24.04	14.61 to 16.46	35 to 54
	Pearl millet	22.94 to 29.28	16.33 to 23.87	23 to 47
4. Central Rice Research Institute, Cuttack (Jha *et al.*, 1983)				
Rainfed upland paddy; Improved variety (CR-22-MW-10), seed rate 80 kg/ha, 35 kg/ha, control of gundhi bug with lebaycid, two hand weeding	Rainfed upland paddy	42.80	9.50	350
5. Central Tuber Research Institute, Trivandrum (Ramanathan *et al.*, 1984)				
Cultivation of cassava hybrid (H.1687 and H.2304), introduction of groundnut as intercrop	Cassava	28.00	12.00	350
	Groundnut	6.80	15.12 to 22.00	74 to 113

Research Institute/ Improved technology	Crop	Improved practice (q/ha)	Farmers' method (q/ha)	% incr-ease over farmers' method
6. National Bureau of Soil Survey and Land Use Planning's Regional Centre, Delhi (Naidu *et al.,* 1983)				
Balance use of chemical fertilizer on the basis of soil test, Paddy – NPk 10:40:40:, Wheat 100:40:40, Pearl millet NPk 80:40:40, improved variety : Paddy (Jaya), Wheat (KSML), Pearl millet (B.J. 104)	Paddy	32.30 to 38.20	15.12 to 22.00	74 to 113
	wheat	31.41 to 53.37	21.35 to 37.10	21 to 46
	Pearl millet	12.57 to 28.00	8.75 to 12.00	21 to 46
7. ICAR Research Complex for North Eastern Region, Shillong (Choudhary *et al.,* 1986)				
Potato : Seed (Kufri Jyoti), proper method of planting, timely fertilizer application, weed control, prophylactic measure for control of insect/ pest	Potato	56.06	36.30	88
	Poulltry	158 egg	66 egg	163
8. ICAR Research Complex and Krishi Vigyan Kendra, Saramsa, Sikkim (Rai *et al.,* 1988)				
High yielding varieties, bala-nced fertilizer, application of furadon granules for control of leaf whorls in maize, spraying demecron for control of leaf roller and stem borer for paddy, application of thimet for control cut worms and red	Rice	25.00	12.50	100

Research Institute/ Improved technology	Crop	Improved practice (q/ha)	Farmers' method (q/ha)	% incr-ease over farmers' method
ants for potato, roughing and disposal of infested infected clones (Rai *et al.*, 1986)				
	Maize	35.00	9.00	311
	Finger millet	20.00	9.00	222
	Soybean	15.00	10.00	50
	Wheat	20.00	10.00	100
	Potato	150.00	60.00	150

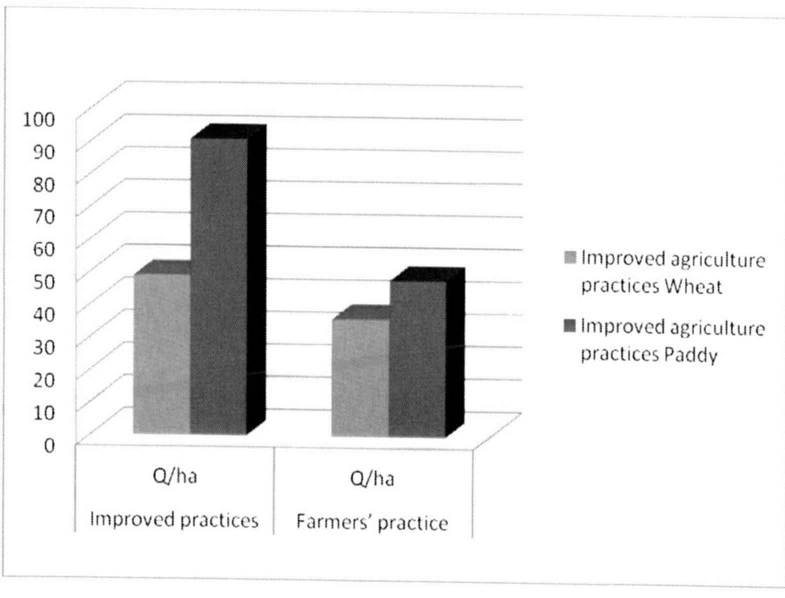

Fig. 5.1a. Improved agriculture practices on crop and its impact on crop yield in LLP villages. Punjab Agricultural University, Ludhiana.

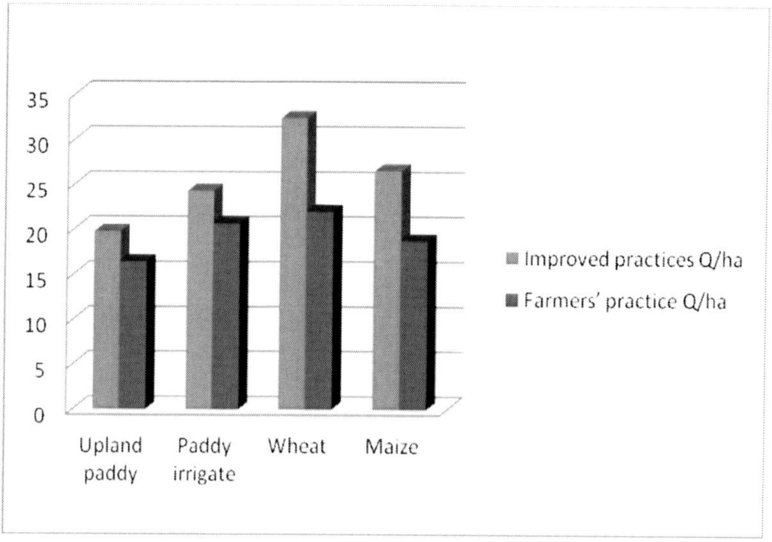

Fig. 5.1b. Improved agriculture practices on crop and its impact on crop yield in LLP villages. Central Soil and Water Conservation Institute, Dehradun.

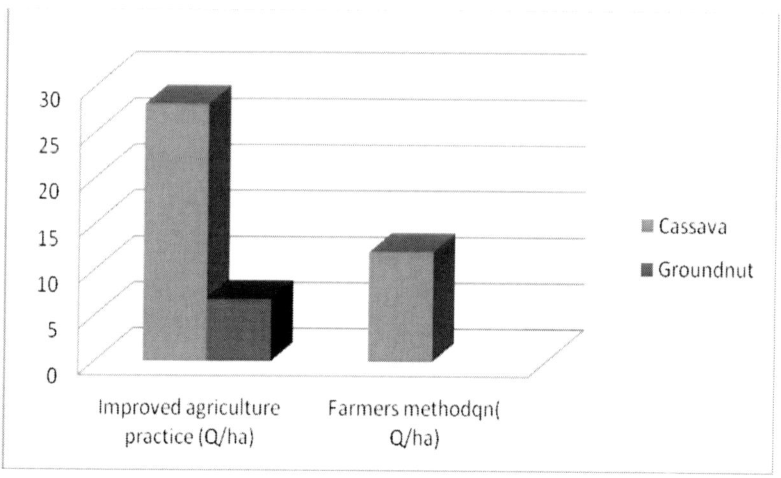

Fig. 5.1c. Improved agriculture practices on crop and its impact on crop yield in LLP villages. Central Tuber Research Institute, Trivandrum.

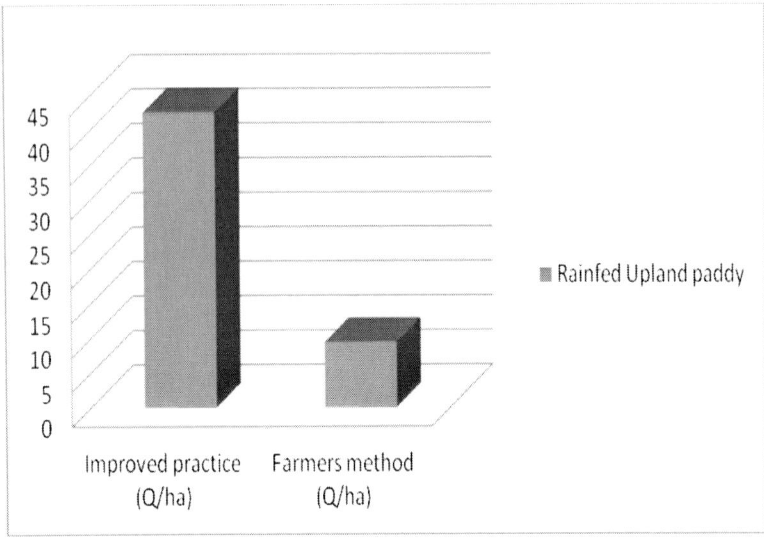

Fig. 5.1d. Improved agriculture practices on crop and its impact on crop yield in LLP villages. Central Rice Research Institute, Cuttack.

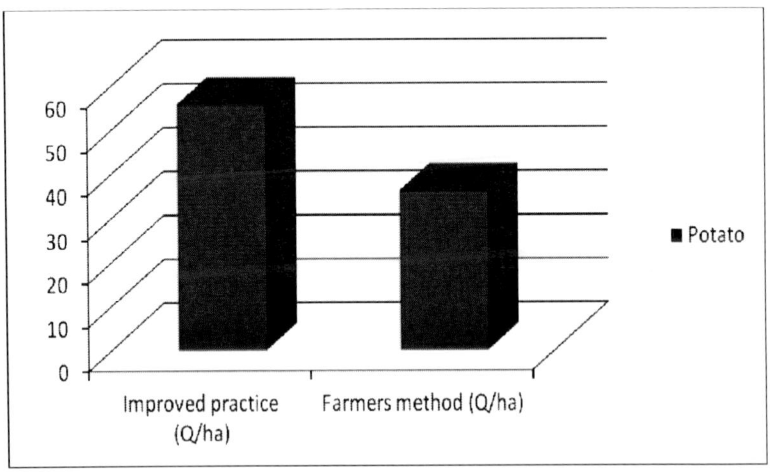

Fig. 5.1e. Improved agriculture practices on crop and its impact on crop yield in LLP villages. ICAR Research Cimpex for North Region, Shillong.

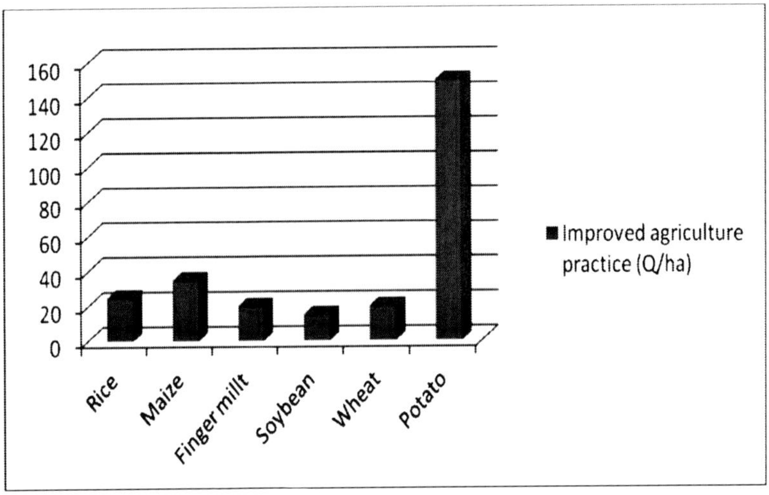

Fig. 5.1f. Improved agriculture practices on crop and its impact on crop yield in LLP villages. ICAR Research Complex and Krishi Vigyan Kendra, Saramsa, Sikkim

Monetary gain from Improved Agriculture Practices to Farmers in LLP

Investigations were made by the scientists to assess monetary gain to the farmers from improved agriculture practice in LLP villages. The income from agriculture practices was calculated by deduction of amount of cost of cultivation from gross income of agriculture practice. The benefit-cost ratio was assessed by dividing gross income by cost of cultivation of agriculture practice. Percentage increase in income and benefit cost was worked out. The detail about this is presented in Table 5.2. The studies reported a significant monetary gain to the farmers from adoption of improved agriculture practices of crop cultivation and animal husbandry as reported by the studies.

There was net profit of Rs. 3312 per hectare in improved method of paddy cultivation while there was a meager returns of Rs. 104.94/ha under farmers' method of cultivation. Improved method of potato cultivation gave net profit of Rs. 2,770/acre as against Rs. 1,470 under farmers' method of practice showing 173.26 per cent increase in net income. Income from scientific method of poultry rearing was Rs. 1119.50 which was 273 per cent more to traditional management of poultry (Rs. 300). The per cent increase in monetary gain from paddy, groundnut and sunflower over farmers method of cultivation was 48.65, 34.46 and 28.45 respectively. Similarly the per cent increase in net income from dairy was 495 per cent more than traditional method dairy management.

The study of **Kumar and Prakash** (1989) revealed a highest return of pea (Rs. 5802) and lowest (Rs. 43) from finger millet. The benefit-cost ratio varies from 1.01 to 4.08. There was a very high benefit cost ratio of pea (4.08), lentil (3.16), amber pop corn (2.41) and wheat (2.00). This shows that by investing one rupee the farmers get back Rs. 4.08 from pea, Rs. 3.16 from lentil, Rs. 2.41 from amber popcorn and Rs. 2.00 from wheat.

The study of **Mishra et** ***al.*** (1986) on demonstration plots of crops in Jabalpur indicated the highest return (Rs. 2,440) from wheat crop and lowest return (Rs. 993) fro sorghum. The benefit cost- ratio varies from 1.52 to 2.00.

Sharma et al. (1986) made a comparison of income from the crops of the beneficiary farmers before (1981-82) and during implementation of the programme (1982-83 and 1983-84). The net return of paddy increases from Rs. 1209 in 1981-82 to Rs. 1,878 in 1983-84 showing an increase of 55 per cent. Similarly, the increase in the return in wheat Rs 2642 in 1981-82 to Rs. 2,851 in 1982-83 due to drought condition, whereas in 1983-84 due to drought condition the net return from wheat crop declined although it was reasonably high (Rs. 2,282).

The study by **Mishra and Jha** (1985) assessed impact of the LLP programme in seven villages of Bihar. All 65 adopted farmers were included in the study and equal number of non – beneficiary farmers was also taken making 130 respondents for the study. The income from the crops of the beneficiary farmers (Rs. 16,995) was also significantly higher than that of non- beneficiary farmers (Rs. 11,910).

It could be concluded that there is enough scope to increase the income of the farmers with adoption of improved agriculture practices. Although there is additional cost in improved seed, chemical fertilizer and plant protection measures in its adoption.

Table 5.2. Monetary gain from improved agriculture practices *vis – a- vis* farmers' practice in LLP

Author	Crop	Monetary gain from improved practice (RS/ha)	Monetary gain from farmers' practice (Rs/ha)	% increase over farmers' practice
Jha *et al.*(1983)	Paddy	3312	104.94	3056
Choudhary *et al.* (1986)	Potato	1,421	520	173.26
	Poultry	1,119	300	273.17
Bheemappa (1992)	Groundnut	7,725	5,780	34.46
	Sunflower	5,395	4,200	28.45
	Paddy	5,625	3,784	48.65
	Dairy	1,690/buffalo	285/buffalo	495

Table 5.3. Monetary gain from improved agriculture practices of crops at farmers' fields in LLP

Author	Crop	Gross income (Rs /ha)	Cost of cultivation (Rs/ ha)	Net income (Rs/ha)	B.C. ratio
Kumar and Prakash (1989)	Finger millet	2,400	2,357.97	42.03	1.01
	Soybean	6,050.00	3,020.00	3,029.60	2.00
	Maize	5.557.85	3,618.54	1,939.31	1.53
	Amber popcorn	8,750	3,618.54	5,131.45	2.41
	Paddy	3,440.00	2,674.87	765.13	1.28
	Wheat	4,379.40	2,323.62	2,055.78	1.88
	Pea	7,680	1,878.33	5,801.67	4.08
	Toria	3,152.50	1,	1,261.46	1.66
	Lentil	5,400.00	1,707.30	3,692.61	3.16
Mishra *et al.* (1986)	Paddy	3,160.82	1,866.35	1,305.03	1.67
	Soybean	2,521.92	1,528.39	993.58	1.64

Author	Crop	Gross income (Rs /ha)	Cost of cultivation (Rs/ ha)	Net income (Rs/ha)	B.C. ratio
	Sorghum	1,470.12	971.83	498.29	1,52
	Redgram	2,475.00	1,349.42	1,126.58	1.82
	Wheat	4,781,77	2,341.92	2,439.91	2.00
	Gram	2,926.16	1,459.48	1,466.69	1.98
	Pea	4,605.36	2,595.80	2,109.55	1,78
	Bhindi	4,782.19	2,443.44	2,338.75	1.95
	Green gram	1,918.90	1,656.44	1087.95	1.80

Section 2
CASE STUDY

Some researchers used case study method to understand how the farmers were educated under the programme and what improvement occurred.

Rao (1982) made case study of a farmer in a village – Chouldary – adopted by Central Agricultural Research Institute (CARI), Port Blair in 1978. Atul Krishna Saojal was one of the beneficiaries of the programme. He had two hectare of land in Andaman. His condition was almost typical of the farmers in Andaman who did cultivation on saline soil. In Andaman, saline soil farmers either do not grow any crop or grow *desi* varieties of paddy that gives extremely poor `yield. In Kharif 1979, the CARI provided Atul Krishna Saojal with salt-resistant paddy (C.S.R.I, 3, 4 and 5) for one *bigha* of land. Other farmers to whom small quantity of salt – resistant seeds of paddy were supplied did not believe that on saline land good yield of paddy could be had. They therefore fed the valuable seeds to their children and ducks. Saojal was provided with a little quantity of fertilizer and pesticide for application to salt-tolerant paddy crop but instead of applying to the new paddy variety, he used them on his own old varieties. Similarly Saojal did not follow the advice of scientists to raise seedlings and transplant them in a row. He broadcast the paddy. However, to his surprise in the first year itself, he got a harvest of 4.50 tonnes of paddy per hectare, which was more than double of his earlier return obtained with conventional varieties. Surprised by the result, Saojal preserved the valuable seeds and covered 3.3 acres of his land in1980 with varieties (C.S.R.I., 3, 4 and 5) which yielded six tonnes of paddy without fertilizer and pesticides. Shri Saojal's success made so much impact on his neighbourers that they wanted to purchase the valuable seeds from him at a premium, which two years ago they spurned

Saojal was also given some Khaki Campbell (K.C.) duck eggs which got hatched with his own desi brooding hens. Next year he crossed them with local ducks in his backyard. He then had a brood of cross-bred K.C ducks which were laying many more eggs than desi ones under similar backyard management conditions. Shri Saojal's success with CARI's technology transfer of only two items, salt-tolerant paddy and K.C. duck made a big dent on the motivation of farmers in Chouldary village who then began to approach the Institute to help them with more such tangible and short term impact technology which could improve their economic condition.

Ram et *al.* (1988) made a case study of farmers who did pioneering work and obtained remarkable improvement in production and agricultural income in the LLP programme undertaken by Regional Research Stations, *Sarkanda* and Bilaspur. These farmers were Nanku Ram Sahu of Sendri village, Balkrishna Kaushik and Lakhan Dheemar of Semettal in Bilaspur. All of them benefitted from the LLP.

Naku Ram Sahu obtained a very high yield of potato (500 q/ha) and paddy (70 q/ha) by adoption of improved agricultural practices. Earlier he used to obtain the yield of 150-200 q/ha of potato and 10-12 q/ha of paddy by adopting local agricultural practices. As he got high yield, he dug a well and fitted a 5 H.P. motor pump there. He also paid up the bank loan. After the rice harvest, he started growing other Rabi crops like brinjal, tomato, chilies and cabbage and obtained high yield of these crops. The cropping intensity went up to 200-300 per cent. The cultivation of important fruit bearing trees was also introduced under the LLP programme.

Balkrishna Kaushik was getting very poor yield of paddy (15 q/ha), wheat (8 q/ha) and moong (2 q/ha) before he was selected under the LLP programme. By adopting improved agricultural practices he obtained high yield of paddy (42 q/ha), wheat (13 q/ha) and greengram (4 q/ha). The per cent increase in the yield was from

62 per cent to 180 per cent. Before joining LLP programme, he was taking one or two crops in a year, but after being trained effectively and by adopting improved agricultural practices, he started getting vegetable crops and increased the cropping intensity up to 200-300 per cent. The yield of vegetables crops were: potato (262 q/ha), brinjal (300-320 q/ha) and tomato (400-450 q/ha). On the advice of scientists, he also applied the blue green algae (10 kg/ha) in paddy after seven days of transplanting and obtained 2.50 q/ha more yield over the yield without blue green algae. He was given training in budding, layering and grafting of mango, guava, lime and other fruits. After getting the training he grew an orchard and he had been giving the training to other farmers.

Lakhan Dheemar used to sow local seed of paddy that gave 21.3 q/ha. But by adoption of improved varieties he got 46.80 q/ha of paddy. He also adopted better crop rotation. Earlier, he used to grow paddy in kharif long with lathyrus, wheat and linseed. After the LLP programme, he started growing gram and paddy in summer season. He took a loan from the bank after deriving good income from higher yields. He had not only paid his loan, but also constructed a pukka house.

Bajaj and Shaikh (1994) have dealt on success story of adoption of poultry farm by Smt. Lonkar, a widowed landless labour of village Alandi Mahatobachi in Pune district. She had a small hut for rearing goats which was converted into a poultry shed by giving white wash to the walls and fixing wire gauges on the front side. The shed was divided into eight compartments by using wire gauge to keep eight batches of broilers at a time. A mixture of rice husk and lime powder was spread on the floor. The utensils for keeping feed and water were prepared from bamboo and aluminum pots respectively. In the beginning, two lanterns were kept in the compartments for providing heat to the chicks. Afterward, arrangement of the electricity was made and bulbs of 60 watt were installed. In this way small poultry shed was prepared locally.

The first batch of 25 day old chicks was kept in the first compartment in the first week and subsequently in every week eight batches were kept in eight compartments. In this way 8 batches of broilers of different age groups were kept in 8 compartments at a time. Out of the eight batches four batches of chicks purchased from

Poona Pearl by the Gramsevak Training Centre and supplied to Smt. Lonkar. Two batches were supplied by the Poona Pearl at free of cost and two batches were purchased by the demonstrator herself.

Scientific feeding practices followed were: (1) Glucose water – immediately after reaching the farm, (2) Maize bran – For 3 to 4 days, (3) Broiler starter – Upto 5 weeks, (4) Broiler finisher – Next 3 weeks. 3.25 kg feed was required for each bird for 8 weeks.

Medicine: The private hatcheries sell a day old chicks of broiler after vaccinating them. To increase the weight and to reduce mortality following medicines were used: (1) Vitamin A – First week in water, (2) Furosol-One gram for 2 to 4 weeks in one liter of water, (3) 3 Nitro - 1 gram - from 5th week onwards in 1 kg. feed.

On an average 2-3 chicks per batch died in the first two weeks, and 22 to 23 birds were sold per week. She sold 23 birds in the nine week and purchased 25 day old chicks from the hatchery. Every week she sold the birds and purchased chicks in order to maintain continuous sale of birds.

In this way she started earning Rs 80 to 85 per week by establishing small broiler unit. Scientists of the Gramsevak Training Centre, Manjri suggested to her to establish a small broiler farm by taking a loan from the bank. She accepted suggestions readily and wanted a loan to start broiler farm. The project outline was prepared for her by the staff of the Centre. The proposal was submitted to the Bank of Baroda for the farmer, in the name of her unemployed brother. After some persuasion by staff of the Centre the loan was sanctioned. After loan she began 'Bhagirath Poultry Farm 'of 1600 broilers in September, 1988. She was selling 200 birds per week with an earning of Rs. 800 per week. Due to continuous efforts, hard work and adoption of low cost technology recommended under the Lab to Land Programme a single woman farm labourer became the owner of a poultry farm and could change her financial status.

SECTION 3
IMPACT OF THE LLP AFTER ITS WITHDRAWAL FROM THE VILLAGES

An evaluative study of the impact of the LLP programme was carried out by **Lata *et al.*** (1989) six years after the completion of the programme in a village in Shivalik region of Himalaya's. They collected data from 45 families. They found that the yield of wheat (123 to 164%) and maize (79 to 100%) considerably increased in demonstration plot compared with the yield obtained before the LLP programme. But the yield after the LLP programme was decreased from 21 to 33 per cent in maize and 14 to 18 per cent in wheat than the yield in the demonstration plot. However, after the LLP programme there was increase in the yield (39 to 100%) in maize and (13 to 164%) in wheat than before the programme. The net return over the base year was 112 per cent for maize and 134 per cent for wheat crop.

The seed rate of wheat had gone up from 68.00 kg/ha to 95.20 kg/ha and that of maize from 12.70 kg/ha to 20.70 kg/ha. The dose of chemical fertilizer in kg/ha by the farmers before the LLP was: 40 N, 8.60 P and 0.50 K for wheat and 20.14 N and 5.80 K for maize crop. After the LLP programme the dose of chemical fertilizer adopted by the farmers was: 65 N, 26 P for wheat and 56 N, 6.82 P for maize crop. Plant protection chemical was not used before the LLP but it was used at about 1.50 kg/ha after the LLP programme.

The most important aspect of the programme was the introduction of cultivated fodder crop. Before the start of the LLP programme, farmers used to send their cattle for grazing in the catchment area. But after the programme of the LLP farmers started getting stored rainwater from earthen dam and started growing berseem also in addition to raising sorghum and pearl millet in the field for feeding cattle. It was found that sorghum yield increased from 141 to 213 q/ha and that of berseem increased from 220 q/ha to 387 q/ha. As the quality of dry fodder increased livestock population increased from 167 to 263. This affected the occupational structure of the families. After the LLP programme 40 per cent of the families started getting income from agriculture and dairy as against 27 per cent before the programme.

There was increase in the assets of farmers with increase in their income from higher crop yields and supplementary income derived from the sale of milk after the LLP programme. Before the LLP programme there was no tractor, television and scooter in the village. There were one thresher 32 harrow/cultivator, five chaff cutter, 11 radios, 22 radio tape recorders and 12 sewing machines. After the

LLP programme there were one tractor, 4 thresher, 40 harrow/ cultivator, 10 bullock carts, 5 television, 7 scooter, 22 table fan, 26 radio, 22 ceiling /table fan 42 bicycle and 24 sewing machine.

The study by **Arya *et al*.** (1990) in Nanda village in Haryana conducted after withdrawal of the LLP programme also reported similar results. The respondents for the study were 50 families. During the programme the yield of maize and wheat was 25.2 q/ha and 33.5 q/ha respectively which was considerably more than those before the programme. The yield of maize and wheat came down by 24. 24 per cent and 29 per cent respectively as compared to the yield obtained during the LLP programme. However, the yields from maize and wheat were 170 and 240 per cent higher than the yield obtained before the programme.

Before the introduction of the LLP programme, these families had to buy 304 quintals of wheat worth Rs 47,120 from outside to meet their requirements. As a result of the adoption of improved agricultural technology, this shortage was not only met, but there was 443 quintals of marketable surplus of wheat worth Rs. 68,665. The marketable surplus of maize was before the programme (1981-82) was 68 quintals. This quantity went up 422 quintals in post-programme period indicating a net increase of 334 quintals worth Rs 53,808. The farmers in rainfed are skeptical about use of optimum quantity seed rate and chemical fertilizer. The demonstrating high yielding variety of maize and wheat and chemical fertilizer convinced the farmers of the benefits of these inputs. Consequently, the seed rate of maize had gone up from 12 to 20 kg/ha and that of wheat from 68 to 98 kg/ha. Also, there was remarkable increase in the adoption of chemical fertilizer for both crops. The use of nitrogen (kg/ha) before the LLP for wheat and maize was 35.30 and 20.14 which went up 58.05 and 69.67 after the LLP programme.

The increase in cropping intensity from 100 per cent to 200 per cent and involvement of more farmers in dairy sector increased employment opportunities in agriculture sector by 28 per cent and in dairy by 30 per cent over the base year 1981-1982.

Section 4
GAIN IN KNOWLEDGE AND ADOPTION OF IMPROVED AGRICULTURE PRACTICES AMONG THE FARMERS IN LLP VILLAGES

Verma *et al.* (1988) examined the extent of knowledge and adoption of summer green gram. The respondents were 100 farmers selected randomly out of 260 farm families in LLP villages undertaken by Indian Institute of Sugarcane Research Institute, Lucknow. The data were collected before the start and after the completion of LLP. The maximum score obtainable for knowledge and adoption was 100. The study revealed that the level farmers' knowledge and adoption of improved practice of summer green gram cultivation increased as a result of LLP. The initial knowledge and adoption of the farmers of improved practices of summer green gram was 38 and 29 respectively while corresponding score after the programme was 58 and 53 The increase in knowledge and adoption level of beneficiary's farmers as a result of LLP was found to be statistically significant. The highest gain in knowledge as well as adoption as a result of the programme was in seed treatment with Rhizobium culture followed by fertilizer management, improved varieties plant protection, water management, seed rate, spacing and timing of sowing. Before LLP programme, the farmers of the village were not acquainted with the fact that treatment of seeds with rhizobium culture and phosphatic fertilizer help in fixing aerial nitrogen into the soil. Under LLP improved seeds of green gram (PS-16, K-851 and Pusa baisakhi) were supplied and their package of practices were demonstrated to educate the farmers. For wider exposure of green gram cultivation, farmers were invited to participate in *kisan gosthis*, farmers' field demonstrations and at the institute. As a result of these, the farmers gained knowledge and were motivated to adopt the improved technology of green gram cultivation. Important factors which motivated the farmers in adoption of green gram technology were 'high profitability' followed by 'suitability after potato harvest', 'maintaining soil fertility', 'nutritive diet', 'low cost of cultivation', 'low labour requirement' and 'low risk'.

Studies were undertaken to assess the knowledge of the farmers and adoption of improved agricultural practices by them who were given advice on improved agricultural practices along with financial assistance (beneficiaries) under the programme and by side of those who were not provided such assistance (non-beneficiary) of the LLP. A resume of the findings of the studies are in the following paragraphs.

The level of knowledge and adoption of improved agriculture practices of beneficiary and non-beneficiary farmers were grouped into low, medium and high knowledge and adoption. It is presented in Table 5.4. The beneficiary farmers had high level of knowledge and adoption of improved agriculture practices of crops than non-beneficiary farmers. Seventy to seventy five per cent beneficiary farmers had moderate level of knowledge and adoption of improved agriculture practice of paddy cultivation and 60 to 65 per cent non-beneficiary farmers had low level of knowledge and adoption of improved agriculture practice of paddy cultivation **Nigam and Rahad** (1991). Similarly 50 to 55 per cent had moderate level of knowledge and adoption of improved agriculture practice of seasamum while 50 to 58 per cent non-beneficiary farmers had low level of knowledge and adoption of improved agriculture practice of seasamum (**Singh and Waris,** 2002).

Table 5.4. Level of knowledge and adoption of improved agriculture practices among beneficiary farmer and non-beneficiary farmers in LLP villages

Author	Crop	Knowledge						Adoption					
		Beneficiary farmer			Non-beneficiary farmer			Beneficiary farmer			Non- beneficiary farmer		
		H	M	L	H	M	L	H	M	L	H	M	L
Nigam and Rahad (1991)	Paddy	15	70	15	10	25	65	15	75	10	10	20	65
Singh and Waris (2002)	Seasa-mum	22	48	30.00	7	35	58	20	55	25	2	48	50

H= High, M= medium, L= Low

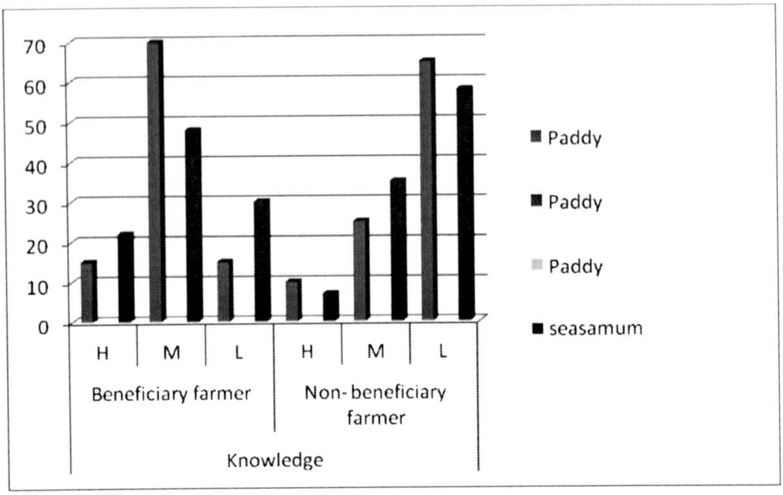

Fig. 5.4a. Level of knowledge and adoption of improved agriculture practices among beneficiary farmer and non-beneficiary farmers in LLP villages

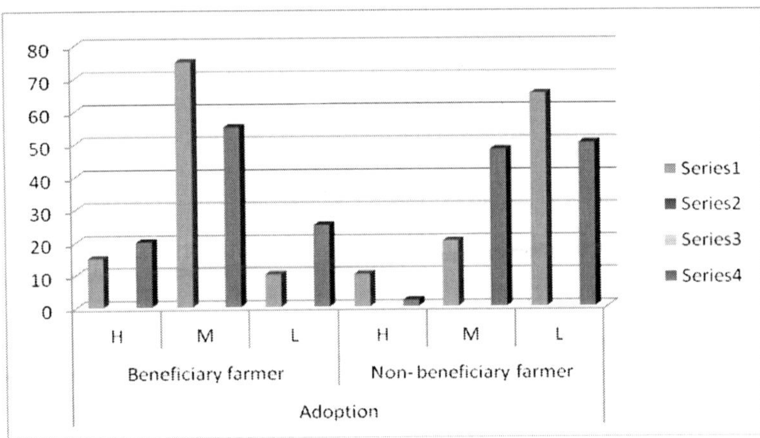

Fig. 5.4b. Level of knowledge and adoption of improved agriculture practices among beneficiary farmer and non-beneficiary farmers in LLP villages

The adoption of improved agriculture practices of paddy and castor has been presented in Table 5.5 shows that the beneficiary farmers had significantly more adoption of improved practices of paddy and castor.

Table 5.5. Mean adoption score of improved agriculture practices of crops among beneficiary and non-beneficiary farmers

Author	Crop	Beneficiary farmer	Non-beneficiary farmer	T value (1%)
Patel and Patel (1997)	Paddy	10.42	7.49	9.502**
Reddy and Reddy (1994)	Castor	44.45	20.93	12.76**

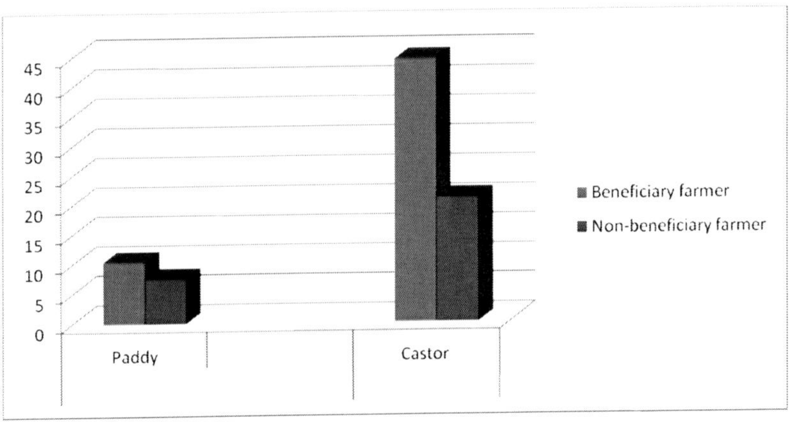

Fig. 5.5. Mean adoption score of improved agriculture practices of crops among beneficiary and non- beneficiary farmers

The study by **Mishra and Jha** (1985) in seven LLP villages of Bihar found that none of the beneficiary farmers had adoption score of improved agricultural practices below 26 and more than half of the beneficiaries (60%) had adoption score in the range of 51 to 75. More than half of non- beneficiary farmers (58%) had adoption score between 26 to 50. The two groups significantly differed with respect to the extent of adoption.

Chand and Sharma (1993) made study to determine the beneficiary farmer's knowledge and adoption level of recommended apple cultivation technology under LLP villages undertaken by Dr. Y.S. Parmar University of Horticulture and Forestry, Solan (Himachal Pradesh). Majority of the beneficiary's farmers had knowledge about the recommended cultural management practices (83%) and establishment of plantation (81%). The farm families' knowledge of training and pruning system, pre and post-harvest technology, fertilizer application and plant protection measures was 67 to 72 per cent. The overall knowledge was 74 per cent.

The adoption of beneficiaries farmers about improved practices apple cultivation was 62 per cent. The adoption of cultural management practices of apple cultivation ranked highest among the beneficiaries farmers (74%) followed by establishment of plantation (70%). The adoption percentage found to be only 40 per cent in case of the use of pre and post harvest technology. The adoption in respect of fertilizer application, plant protection measures and training and pruning ranged Black color between 50 and 65.

The study of Bheemappa (1992) of two villages of LLP of Raichur district showed that the farmers before LLP implementation were using local varieties of seeds and were applying plant protect-ion nutrient less than required quantity in groundnut and more in sunflower and paddy. In case dairy animals, they were feeding of only dry fodder along with occasional green roughage and traditional methods of health care which were not adequate. The demonstration conducted with High Yielding Varieties (HYVs) seeds and recomm-ended plant protection measures convinced the farmers about the benefits of these inputs. Consequently, the HYVs of groundnut (KRG-1, ICGS-11), sunflower (BSH-1, Mahyco Hybrids) and paddy (IR-64) became popular among the farmers. Besides, this, the desired seed rate recommendation in groundnut (137.5 to 150 Kg/ha), in sun-flower (6.25 to 7.50 Kg/ha) and paddy (62.50 to 65.00 Kg/ha) was accepted and was adopted by majority of beneficiaries. With regard to plant protection measures, before the LLP implementation farmers who were not able to identify the pest and diseases and lacked the knowledge of pesticides and their application, became to use recom-mended pesticide to control pest and diseases. In dairying also, the LLP implementation influenced the beneficiaries and others who became aware of the benefits of using concentrates, feeding the cattle mineral mixture, deworming, proper vaccination and health check-up.

Sharma *et al.* (1986) studied the impact of LLP in two villages of Palampur in Himachal Pradesh. The respondents were 30 marginal farming families. The programme was started during 1982-83 and the data pertained to agriculture years 1981-82, 1982 -83 and

1983-84. There was increase of the use of HYV seeds from 71 per cent to 96 per cent during the period in case of wheat. Increase in the HYV seeds of paddy ranged from 4 to 22 per cent. Similarly the use of nitrogen in wheat cultivation increased by 45 per cent, and in paddy the increase was almost 100 per cent during the study period.

Singh's study (1996) reported a wide difference in adoption improved practices between the LLP Village of G.B. Pant University, Ranichauri and a non-adopted village. The adoption of high yielding varieties of crops, viz. wheat, paddy, toria and soybean varied 20 to 80 per cent in the LLP village while it was 10 to 50 per cent in non-adopted village. The improved method of sowing (line sowing) was adopted by a large number of farmers (50 to 60%) of these crops, while in non-adopted village it was far below 9 to 42%. Only a few farmers (LLP village 20%, non-adopted village 9%) had sown mandua crop by line sowing method. The use of chemical fertilizer was encouraging in both villages (LLP village 40-60% and non-adopted village 30%). Both villages were alike in adoption of chemical fertilizer for pea and mandua crops where it was very low (8%). The LLP village was far ahead in application of pesticide to the crops (20 -25%).

Farmers Image of LLP

Patel and Patel (1993) made a study to find out farmers' image of LLP. They collected data from 100 beneficiary farmers and 100 non-beneficiary farmers in 25 village of Valsad and Surat district of Gujarat where the LLP was in operation during the period 1984 -85 to 1985-86. The items included questions to ascertain the correct-ness farmer's image of LLP - its objectives, its activities, its operati-onal method, usefulness, helpfulness of the programme officials, benefits gained by the farmers – in short, their overall impression about the programme. The study revealed that 86 per cent of the beneficiary farmers had a factually correct impression of the programme and 14 per cent had incorrect image about programme. Among the non-beneficiaries, 78 per cent of non-beneficiary farmers did not have correct image about the programme. Only 20 per cent non-beneficiary farmers had factually correct image of the programme. The mean score indicated that the beneficiaries had significantly higher image of the LLP than non-beneficiary farmers. It is important that agricultural scientists should make aware the farming community aware of the various aspects of the LLP. This is an important requirement for making the programme a success among the farming community.

Section 5
IMPLEMENTATION OF LLP

The study by **Mishra *et al*.** (1986) on the LLP programme of Jawaharlal Nehru Krishi Vishwa Vidyala in Shahpura Block of Jabalpur district revealed that out of 346 demonstrations plots, complete package of practices of crops was done in 144 plots (43%), partial package of practices in 175 plots (50%) and none of the practices was done in 22 plots (7%). The farmers' understanding of the content of the training programme indicated that 80 per cent farmers fully understood the content of agronomy followed by the figure 60 per cent for agriculture engineering, 58 per cent for horticulture, 51 for per cent entomology and 40 per cent for plant pathology. The low percentage of farmers who understood the content of plant pathology and entomology was because the names of chemicals were in English which was difficult for the farmers understand and pronounce. More intensive training programmes should be organized for such items with more frequency so that the farmers could fully understand what is delivered to them.

Shri **A.D. Mongia** (1987), Coordinator LLP programme, provided the following suggestions from on his experience of implementation of the LLP programme in Andaman and Nicobar Islands by Central Agricultural Research Institute, Port Blair for vigorous implementation.

1. The scientists adopted villages for two years for the LLP programme. When the programme is completed another two villages are adopted for two years. This is a very short duration for the success programme as the scientists need time to establish a rapport with the farmers. It is therefore essential that there should be a follow-up to the programme in the villages so that the farmers can continue using the latest technology.
2. The new high yielding variety of paddy (IET 4106) was found very suitable for cultivation in the islands. However, the farmers did not keep seed due to lack of storage

facilities. Government department should extend facility for storage to the farmer's facilities. Seed multiplication farm should be established to cater to the local seed requirement.

3. As soon the LLP programme was withdrawn after its tenure of two years, farmers were reluctant to use inorganic fertilizer due to its high cost. A financial agency should extend loan for the purchase of fertilizers.
4. Timely supply of inputs like fertilizer, duck feed, poultry feed etc was a standing problem which obstructed the adoption of new technologies by the farmers. Small poultry and duckery units were successful as there was a heavy demand for eggs in the island. These inputs should be available in time to the farmers.
5. Development of appropriate land use patterns through different cropping systems and diversification of cropping pattern should be included in extension programme. The farmers should be educated about the principle of soil conservation measures to minimize the effect of erosion caused by torrential rains and water harvesting technology. The administration should take up the eradication programme for cattle humsore disease.
6. Small and marginal farmers can not make good living on a single occupation so therefore these farmers have to adopt more than one occupation to generate enough income for better living.

Bheemappa (1992) made the following suggestion to overcome the constraints in the LLP.

1. The amount (Rs 200 per family) sanctioned for supply of the critical inputs was not sufficient to meet the required input expenditure as per recommended technology. Hence, the amount has to be increased sufficiently to meet the recommended input supply so that demonstration becomes a worthwhile exercise and becomes a model for the farmers in the village.
2. Dairy farmers were convinced of demonstration results of feeding the concentrates and mineral mixture of milch

animals. However, they were not able to adopt these techniques due to low economic status. Financial institutions should come forward to extend loan facilities at subsidized rate to these poor farmers for feed concentrate, vaccination etc.

3. There was non-availability of good seed material like KRG-1 in groundnut at right time of owing for Rabi season as the farmers were forced to sell the product of previous season due lack of storage and in many cases indebtedness. State/voluntary and cooperative agencies have to shoulder the responsibility for creating proper seed storage facilities and have a seed multiplication programme to meet the seed demands of farmers.

4. There was no groundnut high yield variety to sow during months of September and October since majority of farmers face water scarcity for Rabi/summer crop. Hence research should be initiated for evolving the groundnut high yielding variety suitable for sowing in September and October.

REFERENCES

Arya, S.L., Mittal, S.P. and Agnihotri, Y., "Lab to Land Programme Helps Boost Crop Yields in Nada Villages", Indian Farming, XL, 6, pp.23-25, 1990.

Bajaj, S.S. and Shaikh. "Lab to Land Programme for Poors", Maharashtra Journal of Extension Education", XIII, pp. 317-319, 1994.

Bheemappa, A., "Appropriate Technology Transfer Under Lab to Land Programme –A Way to Progress", Agricultural Situation, XLVII, 4, pp. 267-270, 1992.

Bhullar, B.S., Sokh, S.S., Bhalla, J.S. and Sooch, B.S., "Lab to Land Programme Succeeds in Udhowal and Mand-Udhowal Villages", Indian Farming, XXXVIII, 4, pp. 23-25, 1988.

Chand, M. and Sharma, D.D., "Knowledge and Adoption of Recommended Apple Cultivation Technology through Lab-to-Land Programme in Himachal Pradesh", Maharashtra Journal Extension Education, XII, pp. 341-344, 1993.

Chandra, Subhash, Sharma, G.C. and Nambiar, K.T.N., "Lab to Land Programme Boosts Production in Gujarat", Indian Farming, 40, 12, pp. 34-36, 1991.

Chaubey, Awadesh Kumar., "A Study of lab to Land Programme in Varasani (U.P.)", Ph.D. Thesis, Department of Extension Education, Institute of Agricultural Sciencies, Banaras Hindu University, Varanasi, pp 1-205, 1988.

Chaubey, Awadesh Kumar., "A Study of Lab to Land programme of Varanasi in U.P., Ph.D. Thesis, Department of Extension Education, Institute of Agricultural Sciences, Banaras Hindu University, Varanasi, pp. 1-205, 1988.

Choudhary, A.N., Das, A., Singh, R.P., Goswami, S.N. and Prasad, R.N., "Lab-to-Land Programme Transforms a Village in Meghalaya", Indian Farming, Indian Farming, XXXVI, 4, pp.16-18, 1986.

Jha, K.P., Kulshrestha, J.P., Mathur, S.C. and Mishra, A.K., "Lab to Land Programme : Upland Rice Responds Well to Good Management", Indian Farming, XXXIII, 1, pp.32-33, 1983.

Kumar, Vikas and Prakash, Ved., "Lab-to-Land Programme Makes an Impact in Uttar Pradesh", Indian Farming, XXXIX, 8, pp. 30-31. 1989.

Lata, Swarn, Agnihotri, Y. and Mittal, S.P.," Impact of Lab to Land Programme in Shivalik foothill Villages", Agricultural Situation, XIIV, 9, pp. 695-699, 1989.

Mishra, B.L., Awasthi, P.K. and Khan, N., "Impact of Lab to Land Programme in Jabalpur", Indian Farming, XXX, 11, pp.5-7, 1986.

Mishra, R.N. and Jha, P.N., "Impact of lab to land Programme in Increasing Adoption, Productivity and Income of Farmers in North Bihar", Indian Journal of Extension Education, XXI, 1 & 2, pp. 77-79, 1985.

Mittal, S.P., Sud, A.D. and Agnihotri, Y., "Lab to Land Programme Makes an Impact on Shivalik foothills Region", Indian Farming, XXXVII, 1, pp. 12, 1987.

Mongia, A.D., "Lab to Land Programme in Andmans", Indian Farming, XXXVI, 10, pp.27-31, 1987.

Naidu, L.G.K., Verma, K.S., Jain, S.P. and Rana, K.P.S., "Bumper Crops Yields through Lab to Land Programme", Indian Farming, XXXIII 10, pp. 37-39,1983.

Nigam, T.R. and Rahad, B.G., "Impact of Lab-to Land Programme in Development of Small and Marginal Tribal farmers", Maharashtra Journal Extension Education, X, 2, pp. 303-304, 1991.

Patel, B.S. and Patel, R.B., "Farmers Image regarding Lab–to–Land Programme in South Gujarat", Maharashtra Journal Extension Education, X II, pp. 229-230, 1993.

Patel, B.S. and Patel, R.B., "Impact of Lab to Land Programme on Adoption Behavior of Beneficiaries and Non- beneficiaries Farmers in Gujarat", Rajasthan Journal of Extension Education, 3, 1, pp. 22-25, 1993.

Rai, R.N., Dhiman, K.R. and Singh, Japal, "Lab to land Programme in Sikkim", Indian Farming, XXXVI, 1 pp. 39-42, 1986.

Rai, S.D. and Dhana Kumar, V.G., "Lab to Land Programme Helps Fish Prawn and saline Water Paddy Farming in Kerala", Indian Farming, XXXVII, 1, pp. 23-25, 1987.

Ram, G., Kashyap, R. and Sharma, P.N.., "Lab to Land Programme Makes Headway in Bilaspur", Indian Farming, XXXVII, 10, pp. 19-21, 1988.

Ramanathan, S., Anantharaman, M. and Gadewar, A.U., "Cassava in Lab to Land Programme", Indian Farming, XXXIII, 12, pp. 68-69, 1984.

Rao, J.R., "Lab to Land Programme: A Success Story from Andaman and Nicobar Islands", Indian Farming, XXXII, 2, pp.23-24, 1982.

Reddy, T. Rayapa, Reddy, S.S. Manmohan, "Adoption of Castor Crop under Lab-to-Land Programme", Maharashtra Journal Extension Education, XIII, pp. 67-70, 1994.

Report of National commission on Agriculture, Part XI, Research, Education and Extension, Govt of India Ministry of Agriculture and irrigation, New Delhi, pp. 130-134, 1976.

Sanoria, V.C. and Sharma, D. K., "Comparative Analysis of Adoption Behavior of Beneficiaries of Farm Development Programmes", Indian Journal of Extension Education, XIX, 1 &2, pp.84-86, 1983.

Sanoria, V.C., Sathawane, R.A. and Agrawal, S.K., "Impact of Farm Technology on Small Farmers under Lab to Land Programme", Maharashtra Journal Extension Education, VII, pp. 127- 132, 1988.

Saxena, D.B., Saxena, A.P. and Dudani, A.T., "ICAR Golden Jubilee : Lab to Land Programme for Transfer of Technology", Indian Farming, XXIX, 3, PP.25-32, 1979.

Sharma, H.R., Sharma, S.K. and Bhardwaj, R.C., "Impact of Lab to Land Programme – A Study of Marginal Farmers", Agricultural Situation, XLI, 7, pp.547-550, 1986.

Singh, Bhagwan and Waris, Amtul, " Farmers' Knowledge and Adoption of Improved Practices of Til Production in Arid Areas of Rajasthan", Rajasthan Journal of Extension Education, 10, pp.104 -107, 2002.

Singh, Bhagwan, "Adoptions of improved agricultural Ponches by little farmer of Uttar Pradesh, Agricultural Extension Review, 8, 1, pp. 10-11, January-February, 1996.

Singh, Room and Singh Mahesh Inder, "Impact of Lab to Land Programme and Marginal Farmers", Indian Farming, XXXIII, 11, pp. 33-35, 1984.

Tyagi, P.C., Agarwal, M.C., Joshi, B.P., Mishra, A.S., Tomar, V.P.S., Singh, Kalyan and Sharda, V.N., "Lab to Land Programme Makes an Impact in Mussoorie Hills", Indian Farming, XXXVIII, 11, pp.7-9, 1989.

Verma, R.P., Lal, J. and Alam, M., "Adoption of Summer Greengram Technology Through Lab-to-Land Programme", Indian Journal of Extension Education, XXIV, 3& 4, pp. 85-87, 1988.

■■■■

CHAPTER - 6

Operational Research Project (ORP)

The purpose and approach of operational research project as defined in the article by **M.S. Swaminathan** (1975) then Director General of Indian Council of Agricultural Research, New Delhi is reproduced in the following paragraphs:

The emerging agricultural technology is of **two major kinds** with regard to the ease of adoption –

In one kind, for example, the new technology of wheat cultivation, the economic benefits derived by a farmer by adopting the technology are not influenced by what his neighboring farmer does or does not do. In other words, the technology is capable of successful individual adoption, in economic terms.

In the other kind, the economic benefits conferred by the technology on the farmer could be greatly influenced by the extent of co-operative action generated on the part of an entire village or watershed community. Rice and cotton cultivation and prevention of disease epidemics in cattle are good examples. Even in Punjab, with one of the finest farming communities in the world, the average yield of cotton is low as compared to countries like the Egypt where cotton is managed co-operatively without any infringement of individual ownership. Pest control and water management in rice are best done co-operatively in a village. In fact, if this can be accomplished, even some of the fertilizer lost through leaching in drainage water can be recycled, by collecting such water in a pond at the lowest point and re-distributing it.

Source: Swaminathan, Whole Village Operational Research Project: Purpose and Approach, Agricultural Situation in India, 30, pp. 333-335, August 1975.

All these aspects of the new technology would need understanding and co-operation among neighboring small farmers. If this is not achieved, the risk element in crop production will be high. Low–cost and low–effort agriculture will be the poor cultivator's response to a high–risk farming situation, unless he has protection by other means such as crop insurance. This is an important reason for relative stagnation of *Kharif* crop production in our country, where in spite of moisture availability; the farmer hesitates to invest in areas which are prone to pest and diseases, water logging and operational problems in land preparation, as in heavy black soils.

Social Aspects

As a result of the considerable amount of work done on land and water management and crop and livestock improvement by our research institutes and agricultural universities, a stage has come when social aspects of rural transformation will have to be coupled effectively with the technological aspects a match between the two is essential for progress. In the past, social scientists usually had an opportunity only to do a postmortem analysis after a project had been declared as not having achieved its purpose. Such and analysis, of course, is valuable for developing future programmes. But the time lost cannot be regained and we are racing against time in generating more opportunities for gainful employment in our villages and in increasing food production. Therefore, the operational problems in transfer of technology should be tackled from the social, administrative, organizational and technical angles concurrently rather than consecutively. It is for this purpose that the working group which formulated the Fifth Plan proposal of the ICAR suggested the initiation of whole village or watershed operational research project. These projects which have been taken up by our agricultural universities and research institutes in collaboration with the extension agencies, banks and other credit institutions, village *panchayats* and other rural organizations, the local administration and above all, the farmers of the area are intended to introduce new land and water use plans for a well-defined area which can help to maximize the economic benefits from ecological strengths of the area and minimize the ecological risks and instability in crop production. The kinds of projects which have been initiated and the areas in which they will be operating are shown below.

	Name of the Operational Research Project (ORP)	Name of the Implementing Agency
1	ORP on Oilseeds, cereals and animal husbandry production in Chittorgarh District, Rajasthan.	Udaipur University, Rajasthan.
2	Integrated milk and crop production for increased productivity, employment and farm income in villages around Karnal.	National Dairy Research Institute, Karnal, Haryana.
3	ORP on Arid land management.	Central Arid Zone Research Institute, Jodhpur, Rajasthan.
4	ORP on Seed potato and other crops grown in areas in Shillong and Patna.	Central Potato Research Institute, Shimla.
5	ORP on Cotton production, cereal and pulses and animal husbandry in Amaravati, Maharashtra.	Punjabrao Krishi Vidyapeeth, Akola.
6	ORP on Groundnut, cereals and pulses on acid soils in Puri district, Orissa.	Orissa University of Agriculture and Technology, Bhubaneshwar.
7	ORP on Jute, Mesta and other agricultural crops and fisheries.	Jute Agricultural Research Institute, Barrack pore.
8	ORP on Reclamation of saline and alkaline soils in District Kapurthala, Punjab.	Punjab Agricultural University, Ludhiana.
9	ORP on Sheep and wool development.	Central Sheep and Wool Research Institute, Avikanagar, Rajasthan.
10	ORP on Reclamation of alkali soils.	Central Soil salinity Research Institute, Karnal.
11	ORP on Integrated land use planning for plantation crops like coconut, palm, and cashew nut areca nuts.	Central Plantation Crops Research Institute, Kasargod, Kerala.
12	ORP on Integrated control of rice pests in Andhra Pradesh, Orissa, West Bengal, Madhya Pradesh and Kerala.	J.N.K.V.V. Jabalpur, C.R.R.I. Cuttack, Deptt of Agriculture, West Bengal; Kerala Agriculture University, A.I.C.R.I.P; Andhra Pradesh. Agriculture University, Hyderabad.

	Name of the Operational Research Project (ORP)	Name of the Implementing Agency
13	ORP on Root-wilt-affected area, Quilon District.	Central Plantation Crops Research Institute, Kasargod, Kerala
14	ORP on Control of white grub in sorghum farming system in Nanded District, Maharashtra.	Marathwada Krishi Vidyapeeth, Parbhani, Maharashtra.
15	ORP on Transforming rural economy through technological changes around Sevagram in Wardha District, Maharashtra.	Punjabrao Krishi Vidyapeeth, Akola, Maharashtra.
16	ORP on Stepping up Crop and fish production in Sundarbans area of West Bengal.	Jute Agricultural Research Institute, Barrackpore, West Bengal.
17	ORP on Maximizing lac production in Chhota Nagpur area, Bihar.	Indian Lac Research Institute, Ranchi, Bihar.
18	ORP on Integrated control of cotton pests to be undertaken in District Ludhiana, Punjab and Coimbatore, Tamil Nadu.	Punjab Agricultural University, Ludhiana and I.A.R.I. Sub – station, Coimbatore.
19	ORP on Livestock and fodder improvement in district Bareilly, Uttar Pradesh.	Indaian Veterinary Research Institute, Izatnagar, Uttar Pradesh.
20	ORP on Composite fish culture.	Central Inland Fisheries Research Institute, Barrackpore, West Bengal.
21	ORP on Rice in the Districts of Cuttack, Orissa Bapatla, Andhra Pradesh, Hooghly and Burdwan, West Bengal.	C.R.R.I. Cuttack, Department of Agriculture, West Bengal, A. P.A U., Hyderabad, Andhra Pradesh.

Social Audit

These projects had a built-in bias for improving the economic well-being of farming families having an annual income of thousand rupees or less. The introduction of a social audit principle in programme formulation is a new experience for agricultural scientists and the lessons learnt would be valuable not only for standardizing extension techniques for small farmers, but also for re-designing experimental approaches where necessary. Another important feature of the operational research project was that after a technical programme has been drawn up by the concerned scientists on the basis of their understanding of the potential of the area, the programme was presented to the farmers of the area where the project was to be undertaken for their comments and criticism. The final project was thus, the outcome of the joint efforts of scientists, extension workers and farmers. Experience had shown that such an exercise would enhance greatly the feasibility of adoption of any innovation; the actual realization of the targets would depend upon the decisions and action of farm families. For example, to produce one million tonnes more of rice the effective collaboration and participation of nearly 4 million farm families may be needed in some of our states. It is the neglect of this human aspect that often leads to disappointments in terms of realizing targets. The Operational Research Projects aimed to reverse this process and develop strategies of production and productivity in an area in the light of farmers' own views and reactions.

Integrated Approach

In the project, the approach to use of the available resources of soil, water, plants, animal and man were to be a total and integrated one. The work was to be done in graded steps. The aim was to involve the entire rural community including children. The project itself was the arena of learning. In other words, work experience could be really made meaningful by using the operational area project as a place where the children can work and learn. For example, in the village covered by the aquaculture project the children of the local school could participate effectively in induced breeding of fishes, in feeding fish in the pond and in understanding the principles of composite fish culture. Similarly, In the lac improvement project in the Chhota Nagpur area, the children of the local schools could participate effectively in all aspect of lac production, tasar silk production and other activities. Learning by doing acquired a real meaning and significance.

Each project was unique and did not have a set of pro forma or programme. The dynamic nature of the biological world was to be fully recognized; hence the success of the project depended upon the extent to which the participants had the mental and operational resilience necessary to achieve growth in an organic world.

RESEARCH STUDIES ON ORP

Sharma (1994) studied the working and success achieved by operational research projects undertaken by research institutes and agriculture universities. His findings are:

National Dairy Research Institute, Karnal – The institute implemented the project of modern technology for milk production under three clusters of villages each having 5-8 villages in the district of Karnal. There was inter-institutional participation in the project. Indian Agricultural Research Institute, National Seeds Corporation and Haryana Agricultural University made seeds of high yielding varieties of grains available for the project area. Similarly, the State Bank of Patiala and Oriental Fire and General Insurance Company assisted the project by providing loans to the farmers and providing insurance cover for high – yielding milch animal respectively.

The cost-and-return analysis of the demonstrations conducted with cereal crops, such as paddy, wheat, and maize were maintained. The data of demonstration conducted on paddy (IR-8), maize (Ganga-5) and wheat (HD–1982) indicated that per rupee return was of the order Rs. 3.11, 1.95 and 3.57 respectively. On an average, the rupee return per hectare of paddy and wheat in the demonstration laid out in the operational area has been Rs. 2265 and Rs. 2209 respectively. The economic analysis of fodder crops introduced in the villages showed that Lucerne + hybrid napier netted highest income i.e. Rs. 3,748 per hectare of all the fodder. This was followed by berseem and mustard with a return of Rs 3,746 per hectare.

The project covered a population of 2764 breedable cows' and 5316 buffaloes. Out of this 44 percent of cows were owned by marginal and small farmers and landless households. Thus, while increasing the productivity of cows and buffaloes, emphasis under the project were to increase income of relatively poor households by making special programmes for them. Together with the production of cross-bred heifers on farmers' fields through artificial insemination, over 100 cows were bought by farmers from National Dairy Research Institute and outside agencies. With the initiation of the project in the area, the milk production potential registered an increase of about 2.50 lakh liters per annum. The crossbred population in the area had registered an increase of 300 percent.

One of the outstanding features of the project was the establishment of Dairy Units which focused attention on employment potential of milk production enterprise either as a constituent of mixed farming system or as an independent source of livehood. It demonstrated through these Units that a family could earn Rs. 700 per cow per year resulting in harnessing of idle and unemployment labour in more productive manner. As a felt need arrangement, Cooperative Milk Marketing was organized in the operational area where collection of milk both in morning and evening from these villages was undertaken involving 90 milk producers. During the year 1976-77, the producers supplied about 890 quintals of milk and received over Rs 1,28,000 as return.

State Bank of Patiala. In a short period of about one and half years, advanced nearly 5 lakhs of rupees for purchase of cross- bred cows, tractors, tube wells equipments, etc. to milk producers.

Central Soil Salinity Research Institute, Karnal: The institute started an Operational Research Project to demonstrate scientific technology for reclamation of alkali soils in farmers' fields in the district of Karnal district in 1975. The project covered an area of about 6000 hectares in three clusters of villages. During the year 1976, 44 demonstrations were conducted in farmers' fields where the essential technology of alkali reclamation was demonstrated. The technology essentially consisted of a package of practices including proper land leveling and bunding, treatment of soil with amendments, selection of suitable crop varieties, raising of rice as first crop in monsoon season, increased application of fertilizers including gypsum, growing of wheat/barley/berseem in *rabi* season followed by *Dhaincha* as green manure crop in summer, proper management and adoption of crop sequence. Average grain yield of rice and wheat in thirty six such demonstrations during the year 1967-77 on farmers' fields had been of the order of fifty one quintals and sixteen quintals per hectare respectively. During the second year, no gypsum was added, and the average yield of paddy and wheat obtained was of the order of 64 q and 20 q/ha respectively. This showed that both rice and wheat could be grown successfully on such soils after necessary management and reclamation practices were followed.

During the year 1977, 146 demonstrations were laid out on such soils at farmers' fields. In addition, horticulture and animal husbandry programmes were also taken up. 100 *ber* plants and 5675 Eucalyptus trees were planted in the operational area.

Punjab Agricultural University, Ludhiana: A similar project as above was initiated by Punjab Agricultural University in the district of Kapurthala covering three groups 2-3 villages each. Bench mark survey conducted showed that the productivity of alkali soils in the operational area was very low, the average yield of paddy being 23.7 q/ha and that of wheat 11.7 q/ha. About 40 such demonstrations were conducted during the year 1976-77 on the farmers' field where the technology for reclamation of such soils was demonstrated. The average yield of paddy and wheat obtained in such demonstrations was 55 q/ha and 16.5 q/ha respectively thereby indicating that the productivity of such soils could be enhanced considerably provided the technology involved by the Central Soil Salinity Research Institute, Karnal could be suitably adopted.

An Operational Research Project on Integrated Pest Control in Cotton was initiated by Punjab Agricultural University in a village of district Faridkot, which was the heart of American cotton growing tract in the State. The total operational area was 2410 acres. Measures were introduced for minimizing carry–over of the cotton pests. In addition, other measures such as, sowing of resistant varieties, timely sowing of crop and judicious use of water were undertaken. They also helped in minimizing the attack of the insect pests on the cotton crop. Weekly observations for monitoring the pest population were made at regular distance of 250 meters. Sprays to control jassid were undertaken in the operational area which resulted in savings of about Rs 65,000 to cotton growers.

As a result of integrated control measures the number of sprayings with pesticides for the control of insect pest in cotton was reduced by 50 per cent.

Central Inland Fisheries Institute, Barrackpore: An Operational Research Project on composite fish culture was initiated by Central Inland Fisheries Institute, Barrackpore at Krishnanagar in the district of Nadia, West Bengal. During the years 1976-78, seventy five demonstrations were arranged in small and large ponds owned by individual fish farmers to demonstrate the technology for increasing fish production. In each ponds four species of fish, namely, Rohu, Katla, Grass Carp and Common Carp were raised. Six training programmes were also conducted for 100 trainees on composite fish culture.

Fish production in the demonstration ponds ranged from 2654 to 4298 kg/ha annually. The production obtained was nearly 800 percent more than the previous average production of 462 kg /ha year from the same ponds. The cost of fish production in such ponds ranged between Rs 2.94 and 3.06 per kg of fish. As a result of this project, the local uneducated unemployed youth of the area grouped themselves together to form three cooperative societies for marketing the harvested fish from the project area. They got some income by way of commission during the harvest period. The West Bengal Government took up extension of this project in a big way for propagating the composite fish culture in the State. Under this scheme, a total of 584 ponds were covered for fish production.

Central Arid Zone Research Institute, Jodhpur: Chauhan *et al.* (1988) conducted a case study of the impact of Operational Research Project on Arid Land Management of Central Arid Zone Research Institute, Jodhpur. The project covered five villages of Mandore Panchayat Samiti in district Jodhpur.

In the arid desert, topography sand dunes wreak havoc from time to time in the form of sand storms. However, if the sand dune is tamed, it can become a big asset. The objective evidence was found by the technology demonstration of sand dune stabilization. The farmer on whose farm it was demonstrated was the owner of 2323 trees on his farm and had considerably increased the number of goat and sheep because of the availability of adequate fodder from these trees. It was estimated that the farmer could have 2018 quintals of fuel, 356 quintals of fodder and in terms of nutrition, 52.12 quintals of protein. The shelterbelt on farm gave many benefits like fuel, fodder, fencing and shelter for animals during the summer season.

A farmer under the guidance of scientists started grass cultivation on his 3.20 acre farm. *Cenchrus ciliaaris* (Dhaman) was sown in 1976 and became available for grazing in 1977. Many advantages were obtained from this. The daily yield of milk per cow increased from 2 liters to 4 liters. He got 16 kgs of ghee every month and sale of which brought him a substantial income. Grass cultivation was even more profitable as compared to pearl millet. Pearl millet gave a net profit of Rs 250 per acre while the grass gave Rs 500, increasing the income by hundred percent.

Another farmer started Ber cultivation with one thousand plants in 1976 and the Ber trees started giving fruits from the year 1978. His farm became a legend in this area. His Ber orchard received wide recognition. To cope with the demand, he started a Ber nursery from which he earned enormous profit. The people from the different parts of India and abroad visited his farm and nursery to see and learn from his Ber cultivation.

The results demonstration on sprinkler and drip method of irrigation were highly encouraging. The sprinkler system irrigated double area from the same quantity of water as compared to flood system. It also saved 12.50 man days per acre for preparation of plots which was required for the floods system of irrigation. The yield of pearl millet shot up by 64% with sprinkler and 153% with drip method as compared to flood irrigation. Under the drip system the same quantity of water extended the irrigated area by 66.6 percent. Since this did not require much labour, the expenditure on wage was also saved.

Before the start of the ORP farmers neither used improved seed nor chemical fertilizer in dryland cultivation. The percentage of farmers using improved seed was 8.6% and the area covered was 6%. Four per cent farmers used chemical fertilizer covering an area of 0.95 per cent. The average use of chemical fertilizer was 9 kgs. With the ORP the use of farmyard manure was also increased considerably from 11% in 1975-76 to 34.75% in 1982-83. In 1975-76 the area under FYM was only 5% which went up 16% in 1982-83.

The demonstration of biogas plant made wide impact in this area. Having seen this demonstration, fifty farmers installed gobar gas plans in their houses. The net profit per year from gobar gas plant was estimated to be Rs 2121 and the investment proved to be highly profitable, i.e. Rs. 6 on one rupee's investment.

It was found that now the age old religious prohibition of not killing rats was losing ground and farmers were gradually individually as well as jointly trying to get rid of this menace. It was reported that a net profit of Rs 2,294 every year was obtained by controlling the menace of rats.

Chauhan *et al*. (1987) study was taken in five villages on ORP undertaken by CAZRI, Jodhpur reported that mean percentage score of knowledge of arid land farming was quite low (39%). This indicated that there is a need for imparting more knowledge with regard to arid land practices to the farmers.

As for adoption, 60 per cent of arid farming practices was adopted by the farmers in the study area. 40 per cent farmers adopted medium level of arid farming practices, 36 per cent farmers belonged to the group of high level of adoption and 23 per cent farmers had low adoption of arid land farming practices.

Among the various training needs, the farmers perceived **'Agronomical practices'** as most important. Similarly. **'Arid Horticulture'** was also an important aspect and was ranked second. Further, Animal Husbandry was another area which was important especially in arid regions which was ranked third. These training needs were followed by Tree Plantation, Nursery Management and Weed Control in order of importance as perceived by the farmers. The training needs which were less important, as perceived by the farmers were Solar Energy appliances/Biogas Plant, drip and Sprinkler system of the irrigation and sand dune stabilization.

Kherde and Prakash (1998) studied the extent of perception of training. The needs of small and marginal farm women in scientific dairy farming practices in ORP area adopted by National Dairy Research Institute, Karnal (Haryana). A sample of 40 farm women in each category having at least one milch animal was selected randomly. The result shows that the small farm women had minimum level of perceived training needs (45%) in feeding practices whereas, maximum level of perceived training needs (78%) was observed in health-care practices, followed by breeding 63 percent and management 51%. The overall level of perceived training needs for all the practices of dairy farming was 59%. On the other hand, the marginal farm had minimum level of perceived training needs i.e. 54%, in management practices followed by 72 and 68 and 56% in health care, breeding and feeding practices, respectively. An overall perceived training needs by marginal farm women in all the dairy farming practices was accounted for 63%. From the above results, it is indicated that the small category of farm women had better level of knowledge than those of marginal category. Overall farm women of both categories had perceived training needs 61% scientific dairy farming practices. Hence, need based skill oriented regular training in modern dairy farming technology should be provided to these categories of farm women with the help of different method so that they can help themselves maintain their animals properly and increase milk productivity to get higher returns and to improve their socio-economic status.

Arya *et al.* (1994) studied adopted and non-adopted villages in two ORPs of which Una ORP was in Himachal and Bunga ORP in Haryana in 1988-89 four years after the start of ORP. The overall impact of the programme in terms of selected indices of development – income, employment, asset generation and productive expenditure – was evaluated through the use of multiple discriminate analyses. It shows the following:

1. Income of the adopted families from agriculture and dairy had increased in addition to increase in their productive expenditure and employment
2. There was significant change in overall economic status of the adopted families differentiating them from those of the non- adopted villages with respect to the selected indices of development.

REFERENCES

Arya, S.I., Agnihotri, Y. and Mittal, S.P., "Impact of Operational Research Project in Shivalik Foothill Village – A Multiple Discriminate Function Approach", Agricultural Situation in India, XLIX, 8, pp.579-581, 1994.

Chauhan, K.N.K., Kavia, Z.D. and Kokate, K.D., "Knowledge, Adoption and Training Needs of the Farmers of Arid Zone", Journal of Extension Research and Education, 23, 4, pp. 236-246, 1987.

Chauhan, K.N.K., Singh, R.N. and Kokate, K.D., Impact of Transfer of Agricultural Technology in ORP Villages (A Case Study) Central Arid Zone Research Institute, Jodhpur, pp. 1-35, 1988.

Kherde, R.L. and Prakash, Om., "Training Needs of Small and Marginal Farm Women in Scientific Dairy Farming", in National Conference – Extension Education For Sustainable Development, Organised By Indian Society of Extension Education at Mumbai, December 29-31. 1998. pp. 4.

Sharma, S.K., "Operational Research Projects : A whole Village Approach", Indian Farming, XXVII, 7 & 8, pp. 117-120, 1978.

Swaminathan, M.S., "Whole Village Operational Research Project: Purpose and Approach", Agricultural Situation in India, 30 : 333-336, 1975.

■■■■

CHAPTER - 7

Krishi Vigyan Kendra: Concept, Structural and Operational Features and Mandates

HISTORICAL BACKGROUND

The **Education Commission (1964-66)** constituted by Government of India recommended vigorous efforts to establish specialized institutions to provide vocational education in agriculture and allied activities at the pre and post-matriculate level to cater to the training need of boys and girls of the rural areas. It suggested the name **Agricultural Polytechnic**. The recommendation was thoroughly discussed by Ministry of Education, Ministry of Agriculture, Planning Commission, ICAR and other the allied institutions during 1966-72. Finally ICAR mooted the idea of establishing innovative institutions called Krishi Vigyan Kendra (Agriculture Science Centre) for imparting vocational training to practicing farmers, school drop-outs and field level extension functionaries.

ICAR's standing committee on Agricultural Education in its meeting held in August, 1973 observed: Establishment of KVKs was of National importance which would help in accelerating agricultural production as well as in improving socio-economic condition of the farming community. Assistance of all related institution should be taken in implementing the scheme.

ICAR constituted a committee in 1973 headed by **Dr. M.S. Mehta** of Sewa Mandir, Udaipur (Rajasthan) for working out a detailed plan for implementing the scheme. The committee submitted its report in 1974.

ICAR institutes, agricultural universities and selected NGOs have been made responsible to operate the KVKs. The first KVK was established in 1974 on pilot basis at Pondichery under the administrative control of Tamil Nadu Agricultural University, Coimbatore. At present there are 634 KVKs established in the country. The number of KVKs in the states have been distributed in eight zones in the country as shown below.

Zone wise KVKs	State wise KVKs
Zone 1. Ludhiana and Punjab (70)	Delhi (1), Haryana (18), Himachal Pradesh (12), Jammu & Kashmir (18), Punjab (20)
Zone 2. North East Region Shillong (83)	Andaman and Nicobar Islands (3), Bihar (38), Jharkhand (24), West Bengal (18)
Zone 3. West Bengal – Calcutta (75)	Assam (22), Arunachal Pradesh (13), Manipur (9), Meghalaya (5), Mizoram (8), Nagaland (9), Sikkim (4), Tripura (4)
Zone 4. Kanpur (81)	Uttar Pradesh (68), Uttarakhand (13)
Zone 5. Hyderabad (78)	Andhra Pradesh (34), Maharashtra (44)
Zone 6. Jodhpur (70)	Rajasthan (42), Gujarat (28)
Zone 7. Jabalpur (100)	Chattisgarh (20), Madhya Pradesh (47)
Zone 8. Bangalore (81)	Karnataka (31), Tamil Nadu (30)

Noumbers in parenthese indicate number of KVKs

Conceptual Changes over a Period of Time

The M.S Mehta Committee (1974) enunciated the following three concepts of Krishi Vigyan Kendra.

1. The Kendra will impart learning through work experience and hence will be concerned with technical literacy. The acquisition of which does not necessary require as pre – condition the ability to read and write.
2. The Kendra will impart training to only those extension workers who are already employed or to the practicing

farmers and fishermen. In other words, the Kendra will cater to the needs of those who are already employed or those who wish to be self-employed.

3. There will be no uniform syllabus for a Kendra. The syllabus and programme for each Kendra will be tailored to the felt needs, natural resources and the potential for agricultural growth in that particular area.

Committee suggested the principles for success of KVKs

1. Accelerating agricultural and allied production in the operational area of the Kendra should be its prime goal.
2. "Teaching by doing" and "learning by doing" should be the principal method imparting skill training.
3. Training efforts should help the poor ones to raise their living conditions.

During the VII Five plan, the mandate of KVK was reviewed and revised as follows:

1. Conducting "On–farm testing" for identifying technologies in terms of locations specific sustainable land use systems.
2. Organising training to update the knowledge of extension personnel to take advantage of emerging advances in agricultural research on regular basis.
3. Organising short–term and long–term vocational training courses in agricultural and allied vocations for the farmers and rural youths with emphasis on "learning by doing" for high production on farm and generating self-employment.
4. Organising frontline demonstration on various crops to generate production data.

The four projects of National Demonstration (ND), Operational Research Project (ORP), Krishi Vigyan Kendra (KVK) and Lab to land Programme (LLP) were implemented by the ICAR, New Delhi. All these four projects were integrated with the KVK during 1992.

In the 11th Five Year Plan an additional mandate was given to KVK. KVK was also to serve as resource and knowledge centre of agricultural technology for supporting initiatives of public, private and voluntary sectors for improving the agricultural economy of the districts.

In order to achieve the above mandates the following broad objectives were set:

Objectives

1. To promptly demonstrate the latest agricultural technology to farmers and extension workers of State Agricultural Department with a view to reduce time lag between technology generation and its adoption.
2. To test and verify the technologies in the socio-economic conditions of the farmers with a view to identify constraints in production and to modify the technologies to make them appropriate.
3. To impart training to the practicing farmers/farm women, rural youths and field level extension functionaries using the method of "teaching by doing" and "learning by doing".
4. To provide back-up, with training and communication support, to the district level departments in their extension programme.

Special feature of KVK

Dr. C. Prasad (1978), Deputy Director General for Extension of ICAR, New Delhi analysed the special features of the KVK as follows.

1. Strong Subject Matter Support: Krishi Vigyan Kendra is an educational innovation. It's one specific feature which would distinguish it from other field level training programmes is the strong back up support by one of the Agricultural Universities or Central Research Institutes and continual infusion of new technology and scientific knowledge resulting in continual renovation of its training programmes.

2. Survey based Courses: In order that the skill-training at Krishi Vigyan Kendra is relevant to the farmers and in-service trainees, a thorough survey would be conducted identifying the needs of training in the villages under jurisdiction. The core staff of the KVKs would conduct the survey themselves under the guidance of a competent economist/extension workers from the supporting Agricultural University/Research Institute. Selectivity principle would operate both for identifying the courses as well as the trainees-courses which could give maximum return and the trainees who are most needy – poor, weaker, marginal and tribal farmers.

*3. **Training by work – experience**:* The prime method of training in the KVK will be through work experience; the trainees will practice by doing what they are expected to do their own farms. Unlike lecture and notebook dominated training courses, it will be farm and workshop dominated training programmes. Here the conduct of training will not start from the lecture rooms to the farm, rather it will start from the farms and the workshop to the discussion assembly. The training programmes will take into account all methods and means which will result into skill development in the trainees. It can be formal, informal and no-formal or a combination of them depending upon the needs and the circumstances.

4. Practical Core Staff: The core staff would be relatively small, but will be supplemented with guest or visiting trainers – practicing farmers, artisans, extension workers and university experts. While the proficiency in practical work would be the principal criteria, the core staff would be of qualified people with rich experience of working with the farmers and with commitment to contribute to rural upliftment.

5. Practical Training Facilities: In order to ensure high quality training KVKs will have scientifically developed farm, fish ponds, well-equipped working laboratories, workshops, poultry unit piggery, goat and other allied facilities consistent with the training needs of the area. The entire unit would be developed as living crop/livestock/fish culture museum. Training would also be imparted on off-campus supervised farms of the trainee or progressive farmers. In addition of the basic requirements, facilities will be developed specific relation to each course to enable it a useful and productive experience for the farmers.

6. Flexible syllabus: The training courses of the Krishi Vigyan Kendras were to be tailored to the immediate needs of the practicing farmers and relatively long-term needs of the farm youth, specially school drop-outs. These have to be determined keeping in view the particular group of trainees–their background, interest, needs, resource and the production potential of the area. At no time the Kendra will have fixed or uniform syllabi; every course will be unique in its content approach, and methodology best suited to the specific training group and the teaching – learning situation.

7. Composite Training : Perhaps for a long time to come only one KVK has to cater to the training needs of the farming community in a district, the Kendra therefore cannot specialize on the basis either of audience (farmers or farm women) or subject – agriculture, horticulture, livestock production etc. The problems of the farmers are so complex, interwoven and interrelated that the specialized approach may not fit as per requirement of the farmers and it may as well create some operational problems. Hence, each KVK will deal with both male and female groups engaged in farming as well as will gradually develop practical facilities for training in all branches of agriculture and allied fields.

In addition, it is expected that eventually they will develop training facilities for home–crafts and cottage industries in collaboration with the Council for Scientific and Industrial Research, New Delhi, Khadi and Village Industries Commission and other concerned organizations. As a matter of fact, the Krishi Vigyan Kendra in time to come should provide training facilities dealing with all aspects of rural life and thus support the Integrated Rural Development movement by way of providing skill-oriented training.

8. Limited jurisdiction: In order to ensure close and continuing contact with the trainees the temptation for wide coverage both in space and numbers has to be avoided. The district should be the unit of operation for a KVK. In fact, even within the district some phasing out of the areas will be necessary for meeting the training needs of the farmers. The thrust in the training is skill development to the point that the farmers can repeat the same on their farms with confidence. In order to attain this objective the area jurisdiction has to be restricted. The National Commission on Agriculture has even recommended reducing this jurisdiction ultimately to one third of the district per KVK.

9. No Certificate/Diploma : The audience of the Krishi Vigyan Kendra will be training the practicing farmers, farm women, fisherman and women, young farmers – boys and girls, agricultural laborers and field level development staff. In no circumstances, the KVK will give any certificate/diploma to the trainers.

10. Follow-up Measures: In the past, training programmes have seriously suffered for want of proper follow – up measures after conducting the training programmes. Therefore, as a follow –up action contact with trainees would be maintained; benefits of trainees would be assessed; need for further and sequential training would be monitored; communication of new knowledge and technology would be done; inputs required and need for other services such as storage, marketing credit, etc. would be ensured, and so on. Linkage with extension services and other development agencies concerned would be developed to strengthen the follow-up measures.

11. Reporting and Linkages: In view of the inter-disciplinary requirements of farming and huge task of training and follow-up, functional linkages among all concerned organizations, institutes and agencies must be developed and maintained. There should be appropriate reporting and communication systems for effective upward, downward and horizontal communication. Thus, there will be a lot of scope to learn from each other's experiences and innovations and keep on improving the training contents and programmes on continuous basis. Training, thus, becomes a dynamic concept.

12. Functional Linkages: In the past, the training of rural people, especially the farmers, has been treated as a casual affair without being given the significance it deserves. On the contrary, rural training is much more complex and difficult than formal educational programmes of the institutions. Since rural development is inter-disciplinary in character, the involvement of varied institutions at various stages is essential.

In the operation and management of KVK, in order to do a successful job of training and the necessary follow–up for technological adoption, functional linkages with sister institutions and agencies is vital. Such productive relationship is beneficial to all concerned agencies. Keeping this consideration as an essential aspect of this scheme, the ICAR has provided only a core staff in KVKs in the disciplines relevant for the area (district) – agronomy, livestock production, fisheries, horticulture, home science, extension education, etc. It is expected that the trainers can be drawn from various other sources/agencies, to meet the practical requirements of the specific vocational courses. Even the farmers having rich experience in a particular course at the KVK could be inducted for training. They will be entitled for all the privileges, honorarium and status that a professor of a university can have for that course.

While appropriate training facilities are being developed on the KVK campuses, the Kendras will also utilize training facilities available at farm, workshops, and developmental programmes sites in the area. This is why each Kendra is expected to have a good survey of the areas/district to locate institutions and facilities which can be utilized in the training programme. A close linkage with the on-going village development work is considered a must for aligning the training programmes to the development requirements of the area. Such linkages would also ensure follow-up actions by the development departments / agencies.

The very management of the Krishi Vigyan Kendra is in the hands of a local Management Committee consisting of a cross – section of development agencies and institutions including the farmers and farm women. This committee meets and reviews the progress made and problems encountered. The committee is expected to devote half of its time to educational programmes of the Kendra.

For scientific back stopping each Kendra is either directly functioning under the administrative control and supervision or they are functionally linked with the nearby Agricultural University or a research institute for effective monitoring of the specialized training programmes in specific discipline.

The Organisation

At the national level authority of KVK is vested in the ICAR. Director General of ICAR is the principal executive which is supported by eight Deputy Director Generals (DDG). Agriculture Extension is made responsible for implementing agricultural extension programme of KVKs in the country.

In view of the effective coordination and monitoring of these KVKs, eight Zonal Coordinating Units (ZCU) located in different parts of the country have been converted to Zonal Project Directorates (ZPDs).The ZPD is headed by Zonal Project Director. Eight ZPD are located the following places: Ludhiana Shillong,

Calcutta, Kanpur, Hyderabad, Jabalpur, Jodhpur and Bangalore. The mandates of the ZPDs are as below.

1. To formulate, implement, monitor and evaluate the activities of Krishi Vigyan Kendra and other projects of the Division of Agricultural Extension of ICAR.
2. To coordinate the works relating to these projects with various agencies such as State Agricultural Universities (SAUs), ICAR institutes, Voluntary agencies, Directorate of Agriculture and Animal Husbandry of the States and other development departments.
3. Formulation of network research projects of technology assessment and refinement in the zone.
4. Coordinating the process of content development and preparation of district specific module.
5. To serve as feedback mechanism from the projects to research and extension systems.
6. To implement the Front Line Demonstrations of oilseeds, pulses and cotton as required by ICAR.
7. To perform other activities needed for implementation of assigned programmes.

A KVK is set up at a district level. The programme coordinator (PC) of the rank of Professor in the field of agricultural extension or Agronomy and six subject matter specialists of the rank of Assistant Professor constitute the faculty of the KVK. A group of technical and other supporting staff is also provided to each KVK. The subject matter specialists are of the following disciplines:

1. Agronomy
2. Plant protection
3. Horticulture
4. Animal husbandry
5. Home science
6. Agriculture extension.

The total staff of the KVK is 16 as below

Post	Number of post
1. Programme coordinator/ senior scientist	1
2. Subject Matter Specialist	6
3. Programme Assistant (Lab technician)	1
4. Programme Assistant (Computer)	1
5. Farm Manager	1
6. Assistant	1
7. Stenographer	1
8. Driver	2
9. Supporting staff	2

Instructional Farm and Building

A KVK has an instructional farm of at least 50 acres for showing agricultural technology to farmers. The demonstration of improved agricultural technology for crops cultivation, vegetable, fruit, poultry and animal husbandry are conducted at instructional farms. The following residential accommodation and buildings are at the KVK.

Items	Plinth area
Administrative cum laboratory	550 Sq. meters
Trainees' hostel	One
Residential quarter	6
Instructional farm	Farm and demonstration unit
Equipment and furniture	Vehicle, farm equipment
Library	Books periodicals

Scientific Advisory Committee

All the KVKs have constituted the Scientific Advisory Committee as per the guidelines of ICAR which comprises of representatives from ICAR institutes, State agriculture university, development departments of the District, media personnel, financial institutions and progressive farmers and farm women apart from the members special invitees also attend the meeting. On an average 16 members and four invitees attend each Scientific Advisory Committee of KVK.

Scientific Advisory Committee meetings are conducted to get necessary guidance and support to carry out the mandated activities of KVK in a more planned and scientific manner. The composition of the committee by the KVKs is furnished below.

Composition	Chairman/ member
1. Head of the host institute	Chairman
2. Director extension of concerned SAU	Member
3. Zonal coordinator of the concerned zone	Member
4. Representative of ICAR institute/ ICAR regional research station if located in the District or nearby	Member
5. Associate Director Research & extension of the zonal research station in which KVK is located	Member
6. District officers of the line departments such as agriculture, horticulture, animal husbandry, soil conservation, social/agro forestry, sericulture, fisheries, irrigation, social welfare, small scale industries etc.	Member
7. Representative of the lead bank of the district.	Member
8. Farm Radio Officer of the AIR in which KVK is located.	Member
9. Representative of Doordarshan centre in which KVK is located.	Member
10. Two representatives of the farmers one small and one big nominated by the Head of the host institute.	Member
11. Two representatives of the farm women nominated by the Head of the host institute.	Member
12. Programme coordinator	Member Secretary

Reporting

Monthly progress report, Quarterly progress report, Seasonal report and Annual Progress reports are prepared by KVKs in time schedule.

The activities of KVK include on–farm testing to identify the location specificity of agricultural technologies under various farming systems, frontline demonstrations to establish its production potentials on farmers' fields, training of farmers to update their knowledge and skills in modern agricultural technologies, and training of extension personnel to orient them in the frontier areas of technology development. As recommended by the Committee of Independent Evaluation and Impact Assessment of KVKs and QRT (2001 to 2004-05), the fourth activities of the KVKs will be to work as resource and knowledge centres of agricultural technology for supporting initiatives of public, private and voluntary sectors for improving the agricultural economy of the district. In addition, in order to speed up the process of dissemination of technology, a large number of various extension activities and production of seeds such as planting materials are taken up by the KVKs.

Chapter - 8

Research Studies on KVK

A review of research studies on training, extension methods, gain in knowledge and adoption of improved agriculture practices among the farmers, agriculture productivity, attitude of personnel of KVK and farmers toward KVK and suggestion for increasing effectiveness of KVK is being given.

PLANNING AND MANAGEMENT OF TRAINING PROGRAMME OF FARMERS

Training given to farmers about agricultural technologies is an important feature of the KVK. The aim of the training programme is to upgrade knowledge and skill of the farmers about scientific agriculture. Scientists conduct research on various components of training to be given to farmers. The following paragraphs deal with the findings of research studies on the various aspects of training programmes.

Courses and Participants

Between 2007-08 and 2013-14, the KVKs organized 2,59,645 courses in the country for 74.04 lakh farmers and farm women. It is significant that farm women constitute 23.97 per cent of the total participants (Dare/ICAR Annual Report). At national level the number of courses and the farmers and farm women who attended during the period under review is given in the following Table 8.1.

Table 8.1. Number of courses held and farmers and farm-women

Year	Number of courses	Number of farmers (Lakhs)	Number of farm-women (Lakhs)	Total (Lakh)
2008-09	35,533	8.07	1.59	9.66
2009-10	31,912	9.59	1.68	11.27
2010-11	40,973	9.00	3.00	12.00
2011-12	43,973	8.77	3.76	12.53
2012-13	51,098	11.44	4.03	15.47
2013-14	48,576	9.42	3.69	13.11
Total	2,59,645	56.29	17.75	74.04

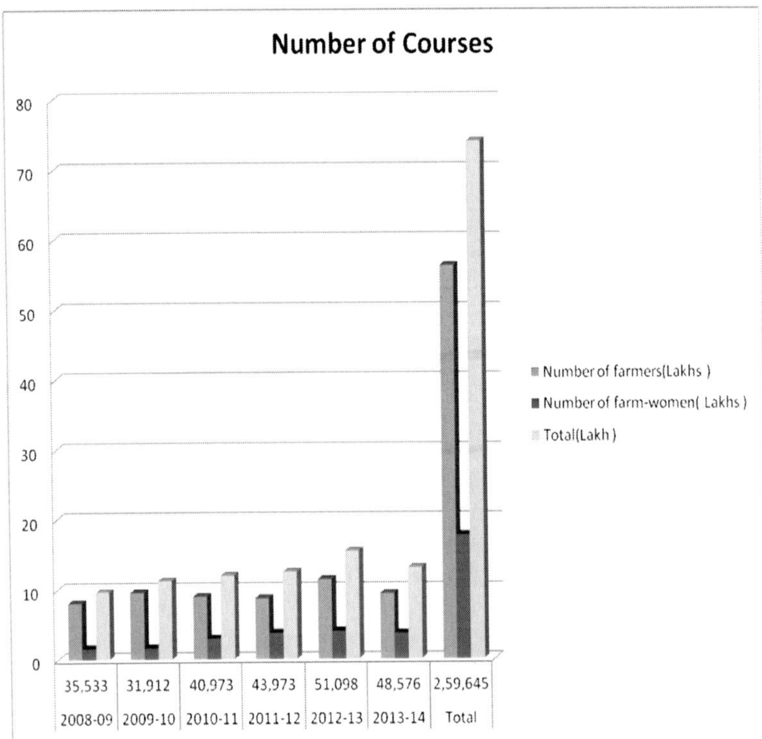

Number of Participants in the training course

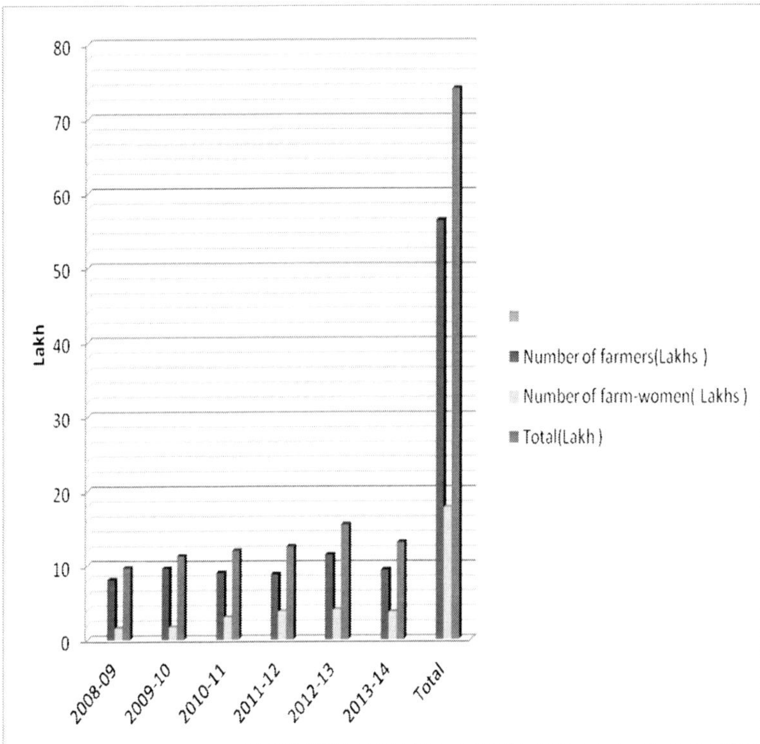

It would be observed that the highest number of courses in a single year was 51,098 in 2012-13 while the lowest number was 35,533 in 2008-09. The average number of courses between 2007-08 and 2013-14 was 43,274. It is reflected in vast increase in the number of farmers and farm women. From 9.66 lakhs farmers and farm-women in the year 2008-09 it went up to 11.27 lakhs in 2009-10, 12.00 lakhs in 2010-11, 12.53 lakhs in 2011-12, 15.47 lakhs in 2012-13 and 13.11 lakhs in 2013-14. It has increased by leaps and bounds in the year 2012-13 and reached an all time of high 15.47 lakhs. The increase in number of courses and participants was mainly due to steady increase in the number of KVKs in the country and also growing interest of the farmer community in acquiring knowledge of scientific agriculture. However, there is a need to increase the number of subject matter specialists and infrastructure facilities of the KVKs to effectively deliver their service to the farmers in the districts.

Subjects Offered in the Courses

The subjects of the course of two years 2013-13 and 2013-14 were in the following Table 8.2. The subjects were mostly similar for the courses during these years. The course covered a wide range of subjects. The largest percentage of courses (12 to 21%) was on crop production, horticulture, plant protection, livestock and empowerment of rural women. Very small percentage of courses (2 to 8%) were on soil health and fertilizer management, fisheries, agro forestry, farm machinery, capacity building and group dynamics and production of farming inputs.

Table 8.2. Percentage of time allotted to courses for farmers and farm-women

Sl No	Courses	2012-13	2013-14
1.	Crop production	22	21
2.	Horticulture	16	14
3.	Empowerment of rural women	13	12
4.	Plant protection	13	13
5.	Livestock production and management	12	17
6.	Soil health and fertility management	5	8
7.	Capacity building and group dynamics	5	5
8.	Production of input at site	4	3
9.	Fisheries	3	3
10.	Agroforestry	3	2
11.	Farm machinery and implement	-	4

Factors Influencing Farmers Participation in Training Course

Nanda and Tantray (1990) studied reasons for low and high participation of women in the training programme from 120 farm women in 12 villages adopted by KVK Malangpora run by Sher-e-Kashmir University of Agricultural Sciences and Technology, Kashmir. There were various reasons for low participation in the KVK training programme. The highest mean score (2.94) was obtained for religious restrictions to go outside the four walls of the house. Another important factor was that most of the respondents had a feeling that it was a sin to attend a male conducted training programme outside the home. The third and the important reason was that the women remained busy in their house hold work and did not find any spare time for attending the training courses. A reasonable majority of the respondents claimed that they were unaware of such activities. However, the least score (1.09) was obtained for the innovativeness of such programme.

Farm women attended training courses for different reasons. Most of the women came to improve their skill in different disciplines. The majority of the women revealed that they improved their skill, liked the exchange of thoughts got rid of domestic work and improved their knowledge.

Training Need

The study of **Ram and Mathur** (1981) in four villages of Kurebhar Block in Sultanpur district of Uttar Pradesh on the need of farmers in cultivation of high yielding variety of wheat reported that the training needs were maximum for small farmers. Weeds and weed control, treatment of wheat and storage of grain had been the main areas of training needs of big farmers and medium farmers. Small farmers needed more training in weeds and weed control, treatment of seeds, and irrigation and water management. They attached greater importance to training in the use of fertilizer and irrigation practices. However, they perceived storage of grain as less important training need as compared to other categories of farmers because the small farmers usually do not have enough grain to store.

The study of **Singh and Gill** (1982) on Divyagyan KVK programmes revealed that farmers had high level training need in wheat cultivation, potato cultivation and farm machinery. In case of wheat cultivation the maximum training need score (0.96) was found in weed control followed by manure and fertilizer requirement (0.94), seed treatment (0.93), plant protection (0.84), selection of wheat variety (0.82), sowing of wheat (0.77) and irrigation (0.75) in descending order. In potato cultivation, the highest training need was found in treatment of potato seed (0.96) followed by plant protection measure (0.95), manure and fertilizer requirement (0.89), sowing of potato (0.86) and selection of potato variety (0.85) in descending order. The average training need scores in different aspect of farm machinery viz., agriculture implement, tractor and diesel engine and plant protection equipment were 0.88, 0.86, and 0.96 respectively. Further it was found that in case of wheat cultivation, potato cultivation and farm machinery comparatively low training need score were in respect of those areas which do not require sophisticated technology and were relatively simpler in nature. On the other hand, the high training need was in those areas which were complicated in nature and required the basic understanding of subject matter.

Khan *et al*. (1993) identified the training need in rice cultivation from 75 farmers in three villages in Kohima district in Nagaland. The farmers perceived 'most essential ' need in the areas - plant protection measures, fertilizer, and manure application, high yielding varieties and use of improved trials and machine. The farmers rated weed control and land reclamation in 'essential' category. The farmers perceived method of transplanting and marketing of paddy as 'not essential' areas of training.

Singh *et al*. (1979) conducted study on training needs of dairy farmers in Karnal district. The respondents were 178 milk producers. A very large number of the respondents (84%) gave no opinion about the subjects to be included in the training programme. This pointed that milk producers were not even aware of the possibility of acquiring new knowledge through training. But among the rest of the respondents there was mixed reactions. Most of the respondents, however, were interested in getting training in the field of artificial insemination and pregnancy diagnosis. As for the feeding practices are concerned, most of the respondents were of the view that they should be imparted in all important aspects of feeding. In case of fodder development majority of the respondents (55%) wanted training on both *kharif* and *rabi* crops. 55 per cent respondents wanted training on cleans milk production and housing of animals. In case of veterinary care 47 percent respondents wanted training in the areas of prophylactic measures and treatment of sterility and common ailments.

Farooqui (1992) on the basis of his study of 102 farm women in KVK Aurangabad reported that farm women were very much desired to know about improved agriculture practice in agriculture, horticulture, animal science, agriculture engineering and home science.

The study of **Jodhale and Bhele** (1998) in KVK Aurangabad district of Maharashtra state reported high need for training of farm women in the areas of vermiculture, solar cooker, biogas plant and smokeless chulla.

Content of Course and Training Methods

Singh *et al.* (2011) made a participant observation of the method of conducting a training course - management of Cows and Buffalo in drought period - organised by Jodhpur KVK from 9 to 11 February, 2010. The training programme was intended to enrich the knowledge of the farmers about improved technology of animal husbandry in a drought year. The syllabus of the training programme covered the following eight subjects: (1) Importance of dairy farming and steps to Increase milk; (2) Control of Endo and Ectoparasite; (3) First Aid; (4) Enrichment of Dry Fodder with Urea; (5) Infertility Treatment; (6) Balance Diet; (7) *Dhaman* (*Cechrus setgerus*); and (8) Infectious Diseases.

The following points of interest emerged from the study of training programme.

(1) Importance of dairy farming and steps to increase milk production: The course methods included lecture and discussion in the class. The trainer reported that the farmers had been maintaining cattle mostly for consumption of milk in the household. He stressed that the farmers should have cattle for commercial purpose also as they had several needs like their children's education, health, marri-age etc. Hence, the farmers should have income from dairy farming. This is possible only when there is more milk production from milch animals. Five essential conditions were described for increasing milk production: (1) good breed, (2) cheap and balanced diet, (3) good health of milch animals, (4) breeding capacity and (5) profitable market facility. The name of the following breeds – *Tharparkar, Rathi* and *Kankraj* of cow and *Murrah* and *Mehasana* of buffalo – that give more milk were communicated to them. The photographs of the breeds of cow and buffalo supplied by National Bureau Genetic Resource (ICAR, New Delhi) were shown to them. The trainees were asked to tell essentials for milk production. They learnt the essentials of milk production.

(2) Control of endo and ectoparasite: The trainees visited the Animal Demonstration and Research Unit at the KVK. They saw *Tharparkar* breed of cows there. Information about the improved animal husbandry practices was communicated to them with the help of details written on the wall there. The trainer explained to them that worms inside the body of animals suck their blood making the animals week and thus giving less milk. They were told that Albendazole tablet should be given three times in a year at interval of four months for the control of these worms. The dose was 1.5 gram tablet for two quintal body weight of the cattle. The tablet was shown to the trainees. Similarly for the control external worms (tick and lice) they were advised to spray cypermethrin (1 ml in 1 liter water). The trainer asked the trainees to name the tablet for the control of ectoparasite. None of them could tell the name of the tablet because it was in English and the word had several letters. The trainer told them that Albendazole had two words 'Al'and 'Benda-zole' . He advised them to pronounce first 'Al'and then 'Bendazole' . This helped them to pronounce Albendazole. The trainer gave a demonstration of the use of Albendazole.

(3) First aid: The training method was lecture and demonstration in the class. The meaning of First aid – treatment for animal before coming to the Veterinary Doctor - was explained. Two medicines – (1) Potassium permanganate (PP) and (2) *Turpentine oil* were shown to the trainees and their uses were explained. They were told that PP was for (a) cleaning wounds, (b) for control of bleeding and (c) for immediate use in case of snake bite. Use of PP was demonstrated and explained to them step wise. Similarly the use of *Tarpin* was demonstrated to the trainees. Most of the trainees were able to tell correctly the uses of these medicines when the trainer asked them.

(4) Enrichment of dry fodder with urea: The training was organized at the Fodder Treatment Yard at the KVK. The subject was explained in two stages. In the first stage, the trainees were told that during drought the animals became weak due to shortage of fodder in the arid area of Rajasthan. In such a situation the dry fodder should be treated with urea. This fodder provides nitrogen to the animals. It lessens lysine of dry fodder and releases energy. At the second stage, the method of treatment of dry fodder with urea was explained and the trainees were given practical training in : (a) weighing four kgs urea and mixing it in 50 liters water, (b) taking 50 kgs dry fodder from the fodder storage and keeping it on the cemented floor, (c) spraying urea mixed water over the dry fodder and (c) covering dry fodder by plastic sheet. They were instructed that urea treated dry fodder should be given in small quantity in the beginning so that animals should first develop the habit of using this fodder. Later the quantity should be increased.

(5) Infertility treatment: The course method included lecture and discussion. Infertility among cows is a major problem in the arid area. The cows should come in heat first at three year age and after three months of calving. However, it does not happen in some cases due to infertility. There is an economic loss of Rs 10,000 to the farmer. The following proves the point : A cow gives on an average five litres milk a day for 200 days. Thus total milk production is 1000 litres. Sale price is Rs 10 for a litre of milk (sale price pertains to the year 2010). Infertility among the cows may be owing to: endoparasitic infection; lack of balanced diet or of mineral mixture of vitamin A and phosphorus. The measures suggested to overcome the infertility were deworming, balanced diet, mineral mixture, injection of vitamin A and phosphorus. The trainer asked the trainees to describe reason of infertility and methods of control of infertility. The trainees gave the reasons of infertility and also described the measures to control it.

(6) Balanced diet: The trainer talked in detail about constituents of balanced diet, namely protein, fat, mineral mixture, Vitamin A, a balanced diet of 7.5 kg having 1/3 dry fodder and 2/3 concentrate for producing 10 litres of milk. A balanced diet is essential for milk production and for conceiving by the cows at proper time.

(7) Dhaman (Cenchrus setigerus) cultivation: The course method was lecture in class room. The trainer discussed the following improved practices: name of grasses - *Dhaman* (*Cenchrus setigerus*), *Motha dhaman (Cenchrus ciliaris)* and *Sewan (Lasiurus sindicus*), the seed rate, distance from plant to plant and row to row, sowing method, yield and protein from grasses. The seed of the grasses were shown to the trainees.

(8) Infectious diseases: The symptom of the following diseases were explained: (a) Enterotoyemia (ET), (b) Black Quarter (BQ), (c) Foot and (d) mouth disease (FMD) and (e) Haemorrhagic Septicemia (HS). The trainees were taught the importance for vaccination for control of these diseases. The vaccination cost Rs 1 for an animal. The treatment of the disease is costly. It is about Rs 1000 per animal. The trainees were advised that they should take their animals to veterinary hospital for vaccination.

(9) Impact of the training on the trainees: The information recalled and new information received by the trainees from the training programme is presented in Table 8.3. It shows that in five subjects out of eight subjects, the percentage of total information recalled and reported as new information is more than 50 per cent, in two cases it is less than 50 per cent and in one case it is less than 30 per cent. Thus, the trainees recalled most of the information in five subjects while moderate in three subjects. The column three of the table shows that information received by the trainees was new for them. This shows that the training has improved the knowledge of the trainees about animal husbandry practices.

Table 8.3. Information recalled and new information gained from the training

Sl. No.	Content of the training programme	Percentage of the content recalled	percentage of the total message received as new information	Ratio of new inform-ation
1	Importance of dairy farming and steps to increase the milk production	47	47	100
2	Control of endo and ecto parasite	52	52	100
3	Enrichment of dry fodder	57	57	100
4	Veterinary first aid	60	60	100
5	Infectious disease	28	23	82
6	Balance diet	61	61	100
7	Dhaman (*Cenchrus setgerus*) cultivation	44	44	100
8	Infertility in milch animal	50	50	100

Utility and Coverage of Content of Training Programme

Jondhale and Sangle (1996) evaluated the training programme of KVK Aurangabad. The respondents were those who attended the training programme in 1992-93. Courses of training one each, in the field of Agronomy, Animal Husbandry and Home Science were selected for the purpose of evaluation. The mean percentage score of utility of content for Agronomy was highest (80%) followed by Animal Husbandry (77%) and Home Science (73%). The mean percentage score of coverage for Agronomy and Animal Husbandry was same (77%) while for the mean percentage score of coverage for Home science was 70%. The mean percentage score of training effectiveness of the training programmes was very high (96 to 100%). Similar finding was reported by Sharma (2002) who evaluated the training programme of KVK in Raigarh district of Madhya Pradesh. 338 trained farmers during 1995-96 to 1999-2000 were the respondents of the study. The mean percentage score of training effectiveness was 93 percent maximum in case of plant protection subjects followed by animal science (89%), crop production (87%) and horticulture (78%). In contrast the training programme conducted by Dhule KVK in 1993-94 was not effectively organized, as reported by Desai *et al.* (1996). The mean percentage score of utility, coverage and training effectiveness of the training programme was 48, 49 and 48 respectively.

The study by **Sharma** (2003) in two KVKs in Punjab revealed that all the farm women were fully satisfied with the course content, message delivery and human relations aspects. Almost all farm women (99%) were fully satisfied on the point of training need.

Soni and Bhimwat (2002) collected data from 50 tribal farm women in Thadol Block of Udaipur district. Majority of the farm women (64%) expressed that the syllabus of the training programme covered most of the things needed. Twenty four percent farm women said that the syllabus provided them opportunity to learn some of the things they needed and 12 per cent of the farm women expressed that the syllabus provided them an opportunity to learn all they wanted.

Saxena and Gour (1999) conducted study to find out the utility of training programme for the trainees in KVK Chhindwara. The respondents were 132 trainees who attended the training programme during the period from 1996-97 to 1997-98. The sample included 40 youths from poultry farming training, 40 women from mushroom cultivation training, 20 youths from dairy management training and 32 youths from fabrication of modern store bins training programme. The study revealed that 45 per cent respondents said that they would use the training programme for self employment; 30 per cent respondents said they would use their training for domestic use and remaining 25 per cent respondents said that they would try to disseminate the training content among farmers.

Physical Facilities for the Training Programme

Sharma (2003) studied the extent of satisfaction of women trainees regarding physical facilities available for a training course in Punjab in two KVKs, namely, KVK Rauni (Patiala) and KVK Abohar. 50 per cent trainees were fully satisfied with physical facilities available while 50 per cent trainees were dissatisfied. All the trainees were dissatisfied with the boarding and lodging facilities available because none of the KVKs provided hostel separately for women. Half of the respondents were fully satisfied with the canteen and drinking water facilities while the remaining half were dissatisfied. As regards equipment and material available for training, 42 per cent trainees were fully satisfied, 8 per cent partially satisfied and 50 per cent trainees were dissatisfied. Eighty seven percent trainees were satisfied with the organization of the course while 13 per cent were partially satisfied.

Sharma and Mahajan (2004) also conducted a survey in all the ten KVKs of Punjab on infrastructure facilities for farm women for the training programme. Data were collected from the Chief Training Organisers, Assistant Professors (Home Science) and Demonstrator (Home Science). The total respondents were 30 in number. KVKs had most of the basic infrastructure facilities except for hostel facility separately for women and the audio-visual laboratory. They lacked some tools and equipments in home science laboratory such as labour saving devices and equipments for fuel energy management. In the audio–visual laboratory most of the KVKs had modern audio–visual equipments, but lacked some essential and commonly used audio–visual aids such as chart, flash cards, photographs, video camera, etc. The KVKs needed modern and advanced equipment like knitting machine, solar cooker, zig-zag embroidery machine etc for advanced training and vehicle for trainees. The researchers concluded that infrastructure and other facilities especially for women oriented training were lacking.

Soni and Bhimawat (2002) studied the effectiveness of training programme for 50 tribal farmer women in Thadol Block of Udaipur district. All the trainees said that lodging and boarding facilities during training programme were adequate and satisfactory. As many as 76 per cent trainees considered the use of audio-visual aids in the training programme as effective. All the trainees said that they were provided enough time for discussion to clear their doubts during the training programme. 78 per cent trainees considered the physical facilities as satisfactory. 80 per cent trainees considered the training programme useful in their gaining knowledge.

Duration and Size of the Trainees

The same study also found that the farm women felt that training should be conducted before commencement of crop season as it will be convenient for farmers to attend the training programme during slack period. 96 per cent of the tribal farm women considered the timing of conducting classes convenient but four percent of them considered the timing not convenient. 90 per cent tribal farm women considered the size of group of 30 participants optimum. Majority of the trainees (60%) expressed the view that three days' duration of training was short.

Farmers' Opinion about Training Programme

The study of **Kumar and Khan** (2005) in Rajasthan revealed that in case of State Agricultural University run KVKs, majority of farmer trainees (80%) had a favorable opinion of the training programme, 20 per cent farmer trainees had most favorable opinion and none of the farmer trainees had least favorable opinion of the training programmes. In case of ICAR run KVKs, 53 per cent farmer trainees had most favorable opinion, 47 per cent farmer trainees had favorable opinion of the training programme and none of the farmer trainees had least favorable opinion of the training programmes. In case of NGO run KVKs, majority of farmer trainees (80%) had least favorable opinion of the training programmes and remaining 20 percent farmer trainees had favorable opinion.

Majority of farmer trainees of SAU and ICAR run KVKs had obtained more favourable opinion of the training programme than the trainees of NGO run KVKs. The reason behind this might be that the SAU - KVKs and ICAR – KVKs were the oldest KVKs having good infrastructural facilities and well structured management system, so their training programmes were found better by the farmers than the NGO run KVKs. Training facilities at NGO run KVKs were at the developing stage.

Impact of Training

Before–after study method was used for evaluation of the training programme to assess the knowledge and skill of the farmers. In this the knowledge of the farmers is assessed before the beginning of the programme and again immediate after the training programme has ended to know how far the farmers gained in knowledge from the training programme. The studies made by researchers on 'Before – after' method was discussed below.

Singh and Das (1985) evaluated of the training programmes: (1) 13 one - day training programme and (2) One month long duration training programme of the KVKs in North – Eastern Region of the country.

From an evaluation of 13 one- day training programme with 184 participants it was found that the existing knowledge about specific aspects of farming before conducting the training programme varied from, as low as, 6.25 to 14.37 which increased from 57.99 to 71. 67 when the programme ended. The knowledge gained, therefore, varied from 44.18 to 64.88 in different training programmes. It is further revealed that the gain in knowledge of the participants in these training programme has been increased on an average from 4.73 to 11.33 times with a total average of 6.80 times, considering all the training programmes together.

From an evaluation of study of 15 participants attending a month long training programme on land development in accordance with the pattern of alternative farming system developed in ICAR Research Complex for North Eastern Hill Region (NEH), it was found that the average knowledge about land development has been increased on an average 6.44 times and the mean knowledge about the land use aspect and method of cultivation has been increased, on an average, 4.54 times and 9.94 times respectively.

The study conducted by **Ramakrishnan and Rathakrishnan** (2013) in three KVKs in Tamil Nadu indicated that 70 per cent trainees acquired medium level of knowledge, 13 per cent trainees high level of knowledge and 17 per cent trainees low level of knowledge from training programme on cultivation of paddy.

Singh *et al.* (2011) evaluated a training programme – Management of Cows and Buffaloes during Drought Period – conducted by KVK, Jodhpur. At the end of instructions of each subject in the training programme, the trainees were asked to recall and tell about the information conveyed to them by the trainer. They were also asked to mention whether the information conveyed to them was new to them. Their answers showed that in five subjects out of eight subjects, the percentage of total information recalled and reported as new information was more than 50 per cent. Thus, the trainees recalled most of the information in five subjects while they could recall the information moderately in three subjects.

Singh and Singh (1995) conducted a study at KVK, Munger (Bihar) to understand to what extent the training in rice technology had been able to inculcate skill in the trainees. Training resulted in increasing the level of skill of the trainees in the experimental group. Both the experimental (mean score 18.00) and control group (mean score 16.94) had almost equal level of skill measured before the training started. The mean score of experimental group and control group after training were 34.93 and 18.00 respectively.

'Ex-post – facto research' method was also adopted to assess the impact of the training programs by the researchers. The findings of studies are given in the following paragraphs.

Ghosh and Pandey (2003) studied impact of training programme on improved rice cultivation techniques in Ranchi district at Divyayan KVK run Ramakrishna Mission and engaged in providing training to mostly tribal poor farmers. 120 trainees and 120 non-trainees were selected from the district making a total sample of 240 respondents. The mean percentage knowledge score of the trainee respondents was 62 and that of untrained respondents was 10. The significant 't' value at 1 per cent level of probability indicated that the knowledge level of the trainee was considerably higher than that of the untrained. Distribution of respondents into different categories of knowledge levels showed that majority of the trainee respondents (62%) had medium level of knowledge followed by 22 per cent high and 16 per cent low level of knowledge whereas all untrained respondents had low level of knowledge in comparison to the trainee respondent.

Veeraiah *et al.* (1988) assessed the knowledge and adoption of 150 farmers of Anantapur district of Andhra Pradesh who were trained at KVK in improved practices of rainfed groundnut cultivation. Majority of the farmers (62%) had medium level of knowledge about recommended practices of rainfed groundnut cultivation followed by low level of knowledge (19%) and high level of knowledge (19%). The possible reason for having medium level of knowledge among majority of farmers could be due to low level of retention of the content of the training passed on to them at the time of training and also because of there were few pre-seasonal training programmes besides, difficulty in understanding the language used in the training which was different from their normal speech, insufficient follow-up and low extension contact of the farmers.

Half of the trained farmers had medium level of adoption of rainfed groundnut cultivation techniques. The percentage of farmers having low level of adoption and high level of adoption were 29 and 21 respectively. The possible reason for this trend might be lack of awareness of improved cultivation practices, high cost of inputs and unfavorable seasonal condition.

Kumbhare and Khonde (2009) conducted a study in seven villages of Selsura KVK in Wardha district. The data were collected from 150 farmers who received training at the KVK on improved practices of crop husbandry and horticulture. The study revealed that 66 per cent respondents gained knowledge of improved practices at medium level, whereas 20 per cent and 14 per cent respondents gained knowledge after training at low and high level respectively. The study also revealed that half of the farmers (50%) adopted the recommendation of improved practices at medium level, whereas the other half of the sample population fell into the categories of low and high level with equal percentage, i.e., 25 per cent each.

KVK Kurukshetra conducted the vocational training on vermicompost, detergent making, bee keeping, and preparation of milk product, fruit and vegetable preservation and mushroom production. Learning of new skills led to adoption of occupations by the trainees. Adoption percentage was – vermicompost (12%), fruit and vegetable preservation (8%), bee keeping (13%), white button mushroom (8%). The trainees ran their respective enterprise very successfully and earned good profit (Chauhan *et al.*, 2006)

Mangat and Gill (1999) examined the impact on the beneficiaries of five training courses at Gurdaspur KVK in Punjab organised during 1991-92. The data were collected from 85 respondents after two years of completion of the courses. Four trainees adopted dairying after the training course on the subject at the KVK. They did well and soon acquired 26 milch animals. The occupation generated employment for seven family members and four hired labourers and generated income of Rs. 3350 per milch animal. Five trainees were already involved in this subsidiary occupation. The number of milch animals increased from 21 to 25 in case of those who had already adopted dairying. As a result of which they created employment for nine family members and six hired labourers. The income of these respondents increased from Rs. 2775 to Rs. 3254 per milch animal after the training course. However, 21 respondents did not start the subsidiary occupation due to various constraints faced by them.

Twenty respondents participated in a training course on poultry. Out of this only two respondents started a new poultry unit with 850 birds. This generated employment for one family member and two hired labourers. The income was Rs. 26 per bird. Three respondents were already engaged in this occupation. The number of birds with them increased from 1650 to 2050 after the training course. This also created employment for seven persons two of whom were of two family members and five hired labourers.

Six trainees of bee training course adopted this occupation after the training with 40 boxes. They created employment for five more persons consisting of two family members and three hired labourers. The income earned per box turned out to be Rs. 472 per annum. The number of boxes with nine respondents who were already in this line has increased from 70 to 73. The training course helped to create employment for three family members and two hired labourers. The income of these trainees before the training was Rs. 416 per box which increased to Rs. 451 per box after the training. Ten respondents did not start bee keeping occupation

Out of five respondents of the fruits and vegetables production course three already were fruit and vegetables growers as a subsidiary occupation while one respondent started it afresh. The new adopter grew fruit (mango) on 0.80 hectares of land and helped to create additional employment for two hired–labourers. The area under fruits and vegetables also increased in case of those who had already begun this occupation from 2.2 to 2.6 hectares, thereby creating additional employment for five family members and eight hired laborers. The income so generated increased from Rs. 36,250 to Rs. 38,727 per hectare per annum after the training course.

The respondents who participated in the fruit plant propagation and nursery course were already engaged in it as a subsidiary occupation. Though the area under production remained unchanged, they created additional employment for one family member and one hired labourer. The income per hectare before training was Rs. 23,500 which increased to Rs 25,625 after the training.

According to **Bari** (1987) GKVK scientists at Gokhale Education Society' Krishi Vigyan Kendra Kosbad Hill, Thane in Maharashtra visited Jamsar village in Jawahar Tehsil of Thane district for village survey to formulate need based training. They found most of the women in the village had no work. Men folk used to go to other areas in search of temporary work. It was a challenge for the scientists to recommend a low cost technology which could give employment to the needy people. The knowledge of the scientists was not useful to solve the problems of the village. During the discussion with one of the enthusiastic tribal women, it was disclosed that in cities papad were being prepared from nagli (a main grain crop of the village) which was very nutritive and fetched good price in the cities. None of the scientists, including home scientists, were familiar with the product and the technology. It was concluded from the discussion that village women could get employment from their home produce, if they were trained in papad making. Natural resources like agricultural produces and availability of labour can be linked with employment. However, a problem was of a trainer who could demonstrate the technology. Two women in Jawahar city were familiar with the technique of papad making, which were made available as resource personnel for papad making course.

Jamsar village was small and consisted of five vadies (sub-villages), scattered at different places. The training courses of five days duration in papad making for 20 tribal women at each vadi was organized to train hundred women for self – employment. Untrained women were trained by trained experts in the village and thus the process was started. Nagli as grain is sold at Rs 1.50 per kilogram; but papad prepared from one kg nagli cost Rs 12. Extra expenditure incurred on the ingredients for one kg papad was not more than Rs 2. It is possible to for a worker to prepare papad from 1 to 1.5 kg nagli during the eight hour period and earn Rs 9.00 to Rs 12.00 per day. The village women in Jamsar, engaged in this industry formed a cooperative society and got it registered to gear up the activity. During the financial year 1985-86, the society sold papad worth Rs 30,000 in Bombay and Pune. The Tribal Development Officer of the Tehsil encouraged the women by linking the society with the tribal residential school for the sale of the product.

The study of **Singh and Das** (1985) in Nagaland about the impact of a home science training programme indicated that among the 70 farm women learned cooking skills 42 per cent made use of the acquired skill. 40 per cent out of the 50 participants in training programme on fruit and vegetables preparation and ten percent out of 50 in tailoring used their skill at home. The study was made after six months of the completion of the training programme.

Trainee's Profile

The study by **Kumar and Singh** (1994) of 100 trainers of KVKs in three states Bihar, West Bengal and Uttar Pradesh revealed that majority of KVK trainers were of the age group 25-38 years. Their cadre was of training associates and they had post – graduation qualification with rural agricultural family background. Most of them had 1-12 years of total service experience, training as first job preference and no exposure to any in-service training. Majority had been involved with seminars and production of folders as professional activity and used discussion mostly for seeking information for professional use. Service to rural people was indicated by large majority as desirable work value and recognition as most stimulating motivation for working more. Majority were slightly satisfied with job. Satisfaction from individual job-items was maximum from status, respect and prestige as a person and least from opportunity to have future promotions. Thus there is need to plan ways to enhance quality of the precious human resource manning the KVK, through modern techniques of job-enrichment, such as rewards, incentives, motivating work environment.

Prasad *et al*. (1999) studied to find out influence of psychological factors influencing quality of working life in KVKs. The study was conducted in eight KVKs in Bihar and Jharkhand. All the 40 scientists and 24 technical cadre personnel were respondents of the study. The job involvement, achievement motivation and organization commitment were psychological variable independents variables and quality of working life dependent variable. It was found that job involvement and achievement motivation were significantly affecting the quality of working life in KVKS.

Section 2
EXTENSION PROGRAMME IN KVKs VILLAGES

Extension Activities

KVKs are mandated to adopt villages for implementation of agriculture technologies. Extension activities are used by them to upgrade farmers' knowledge about agricultural technologies. The following is the account of the efforts and programmes undertaken by KVKs in the villages during the period from 2007-08 to 2013-14.

KVKs use various extension activities to educate the farmers about scientific agriculture. In the period under review 21,26,475 extension activities were organized benefiting 674.71 lakh farmers and extension workers. The average number of extension activities under the review period was 3,54,412 for farmers and extension workers. The extension activities were advisory services, diagnostic and clinic services, celebration of important days, exhibition, exposure visits, ex-trainees meetings, farmer science club conveners meet, farmers' seminar, farmers' visits to KVK, field days, film shows, group meetings, farmers fairs, technology weeks, lectures delivered as resource persons, *mahila mandals* conveners meetings, method demonstrations, plant/animal health camps, scientists' visits to farmer' fields, self help group meetings, soil health camps, soil test campaigns.

Table 8.4. Extension activities organized in the KVKs villages and farmers and extension workers benefited

Sl	Year	Number of extension activities	Number of farmers and extension workers benefited (Lakh)
1.	2008 -2009	2,64,485	80.69
2.	2009-10	3,04,235	106.85
3.	2010-11	2,90,323	106.30
4.	2011-12	3,57,432	108.30
5.	2012-13	4,86,00	170.16
6.	2013-14	4,24,000	102.41
	Total	21,26,475	674.71
	Average	3,54,412	11.24

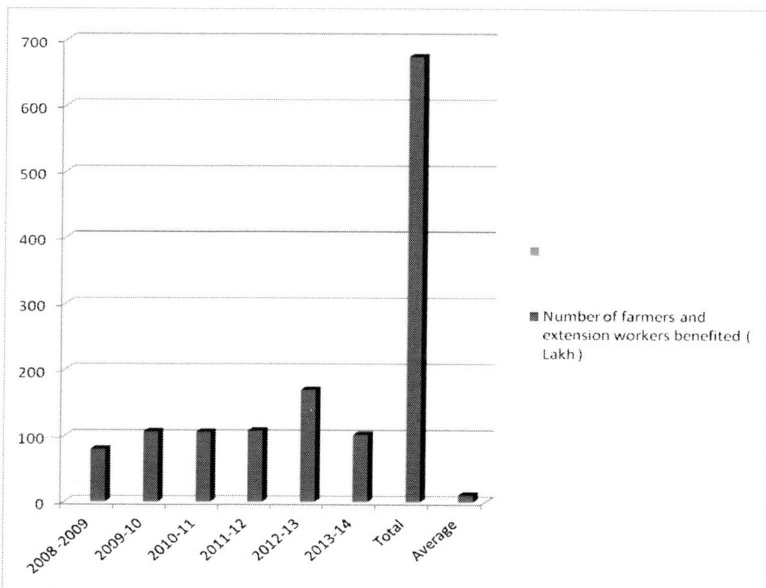

Between 2008 and 2014, the KVKs also organized 6,90,888 extension activities through electronic and print media to have wider coverage of agriculture technology among the farming community and extension workers. The average number of extension activities between the review period was 1,15,158. These included those that used electronic media in the form of television programme, radio talk, CDs/DVDs, and also extension literature, newspaper coverage, popular articles, leaflets, folder and books/booklet.

Table 8.5. Extension activities organized by electronic and print media

Sl.No.	Year	Number of extension activities organized by electronic and print media
1.	2008-09	28,224
2.	2009-10	64,643
3.	2010-11	1,58,464
4.	2011-12	1,69,557
5.	2012-13	1,30,000
6.	2013-14	1,40,000
	Total	6,90,888
	Average	1,15,158

KNOWLEDGE OF IMPROVED AGRICULTURE PRACTICES AMONG THE FARMERS

The scientists conduct research in the field of specialization and make recommendation to be adopted by the farmers. They are to be put into one packet together that is they are listed with details in the form of a pamphlet, bulletin or in the book form depending on the need of the area. This is called package of practices and it is published by the state agriculture university or by the department of agriculture, horticulture, animal husbandry separately or in one packet.

Before finalization of the package of practices, they are first tested under farmers' field condition and are discussed in the zonal and or state level workshops where the scientists and the concerned field officers participate and discuss on each technology to be included in the package of practices. Only the approved package of practices of workshop is published to be adopted by the farmers.

The knowledge of package of practices of agriculture among the farmers is essential for them to adopt these practices in their fields. Several researchers assessed the impact of the KVK in terms of gain of knowledge of improved agriculture practices by the farmers. For this purpose they conducted the study in adopted villages of KVKs and non–adopted villages as control village and selected beneficiaries farmers (BFs), that is, who were given training and demonstration under the programme in adopted villages and those who were not provided such assistance - non-beneficiary farmers – (NBFs) in non-adopted villages. They examined the knowledge of improved agriculture practices of these two categories of farmers of major crops. Their knowledge was assessed with respect to improved agriculture practices of crops-field preparation, improved variety of seeds, seed treatment, sowing method, seed rate, manure, chemical fertilizer, plant protection, post harvesting and storage of crops. The knowledge of the farmers regarding improved agriculture practices of different crops is presented below.

Knowledge of Individual Components of Package of Improved Agriculture Practices

Paddy : Singh (1994) conducted study in Borwat KVK of Banswara district in Rajasthan. The average knowledge of the BFs of high yielding varieties was 51 per cent which was 27 per cent more than the NBFs. However, the gain in knowledge was not substantial. This knowledge can be treated as reasonably satisfactory since improved variety of any crop is highly significant factor for higher production. Similarly, the farmers had poor knowledge about irrigation method of paddy cultivation as the knowledge of the BFs and the NBFs was 64 and 26 per cent respectively. The knowledge of the BFs regarding fertilizer use was found to be high (72%) while it was low for the NBFs (40%). The knowledge level of both categories of farmers with respect to plant protection measures was miserably poor. It was 11 for the BFs and 8 percent for the NBFs. Kharif paddy crop is subjected to greater incidence of pest and disease because of humidity and temperature so the loss in production is very substantial. Under the circumstances adequate knowledge of plant protection measure is very important to minimize production losses in paddy. The farmers' knowledge about cultural practices of paddy cultivation appeared to be only moderate. The BFs had percentage score of 58.00 while it was 40 for the NBFs. Several of the cultural practices are non-monitory inputs for produc-tion, for example – spacing, depth of sowing, time of sowing, age of seedlings. These do not require extra expenditure to carry them out on scientific line, but substantially increase the production of the crop. Special attention needs to be paid to non- monitory factors of production.

Table 8.6. Knowledge scores percentage of improved agriculture practices of paddy cultivation among Beneficiary farmers and Non- beneficiary farmers of Borwat KVK, Rajasthan

S. No.	Improved agriculture practices of paddy	Beneficiary farmers (N=60) Knowledge Score	Non-bene-ficiary farmers (N=60) Knowledge Score
1.	High yielding variety	51	24
2.	Cultural practices	58	40
3.	Fertilizer use	72	40
4.	Irrigation	64	26
5.	Plant protection measure	11	8

Mustard: Singh (1996) assessed the knowledge of the farmers of 11 improved agriculture practices recommended for mustard crop which is a major crop in Bharatpur district in Rajasthan. The study was conducted in Kumher KVK. The BFs had a very high knowledge (86%) of high yielding varieties while the NBFs had moderate know-ledge (62%). The BFs had maximum knowledge of seed rate and recommended spacing (91%) while NBFs had average knowledge (59%). With regard to soil and soil treatment, seed treatment, time of sowing and fertilizer the BFs had good knowledge score (66 to 85 per cent) while the NBFs obtained a score of 38 to 48 per cent for these practices.

The knowledge level regarding weed management and harve-sting and storage was low (44 and 45%) for the BFs and lower for the NBFs (29 and 30%). The knowledge of irrigation management (20%) and plant protection measure (53%) was very poor among the NFS while it was quite substantial (63 to 71%) among the BFs.

Table 8.7. Knowledge scores percentage of improved agriculture practices of mustard cultivation among beneficiary farmers and non-beneficiary farmers of Fatehpur KVK

Sl.No.	Improved agriculture practices of mustard	Beneficiary farmers (BFS) - N=50 Knowledge Score	Non-beneficiary farmers (NBFS) N=50 Knowledge Score
1.	High yielding varieties	86	62
2.	Soil treatment and field preparation	83	49
3.	Seed treatment	66	38
4.	Time of sowing	85	48
5.	Seed rate and spacing	91	59
6.	Fertilizer application	77	38
7.	Irrigation management	63	42
8.	Weed management	45	29
9.	Plant protection measure	71	53
10.	Harvesting and storage	44	30

Another study on assessment of knowledge of improved agriculture practices of mustard crop was by Asiwal *et al.* (2008) in Laxmangarh Panchayat Samiti of Sikar district of Rajasthan. Both types of farmers had maximum knowledge regarding irrigation management (BFs 94% and NBFs 80%) and the time of sowing (BFs 92% and NBFs 79%) and low knowledge of weed management (BFs 45% and NBFs 34%) in the mustard crop. The knowledge of the BFs regarding other aspects like seed treatment, seed rate and spacing, high yielding variety harvesting, threshing and storage, manure and fertilizer application, soil treatment and field preparation, plant protection measure and physiological aspects was 93, 90, 86, 84, 76, 73, 68 and 62 per cent respectively. In case of NBFs the knowledge score was 65, 68, 64, 84, 63, 57, 50 and 52 per cent respectively. There was significant difference in knowledge level with regard to all eleven practices of mustard cultivation among the BFs and the NBFs.

Table 8.8. Knowledge scores percentage of improved agriculture practices of mustard cultivation among beneficiary farmer and non-beneficiary farmer of Kumar KVK in Rajasthan

Sl.No.	Improved agriculture practices of mustard	Beneficiary farmers (BFs) N-60 Knowledge scores	Non-beneficiary farmer (NBFs) N-60 Knowledge scores
1.	High yielding varieties	86	64
2.	Soil and soil treatment	73	57
3.	Seed treatment	93	65
4.	Time of sowing	92	79
5.	seed rate and spacing	90	68
6.	Fertilizer application	76	63
7.	Weed management	45	34
8.	Irrigation management	94	80
9.	Plant protection measure	67	50
10.	Harvesting and storage	84	62
11.	Physiological aspect	62	52

Castor: Kumawat (2008) conducted a study in the KVK Keshwana, Jalore in Rajasthan to assess the knowledge of farmers about improved agriculture practice of castor. The BFs had very high knowledge (64 to 96%) of nine practices out of ten practices. The one practice was that of application of manure and fertilizer (64%). The NBFS had least knowledge about seed treatment (31%), spacing (39%), manure and fertilizer management (27%) and plant protection measures (22%); very high level of knowledge of seed rate (67%) and sowing time (76%) and moderate knowledge of high yielding variety (56%), weed management (59%), irrigation (51%) and field preparation (48%) for castor crop.

Table 8.9. Knowledge score percentage of improved agriculture practices of castor cultivation among beneficiary farmers and non-beneficiary farmers of Keshwana KVK

S.No.	Improved agriculture practices of castor	Beneficiary farmers (N=90) Knowledge score	Non-beneficiary farmers (N=90) Knowledge score
1.	High yielding varieties	81	56
2.	Soil treatment and field preparation	72	48
3.	Sowing time	94	76
4.	Seed rate	96	67
5.	Seed treatment	75	31
6.	Spacing and sowing depth	79	39
7.	Manure and fertilizer	64	27
8.	irrigation management	82	51
9.	Weed management	96	59
10.	Plant protection measure	73	22
11.	Harvesting	86	39

Henna: Meena and Singh (2010) studied the knowledge of farmers growing henna – an economically important commercial crop in Pali district of the arid area of Rajasthan. The 14 improved practices of henna cultivation were selected for the study. The BFs possessed maximum knowledge about high yielding varieties (95%) of henna while minimum about plant protection (54%). Further, soil and field preparation, recommended seed rate, plant to plant distance, time of transplanting, nursery preparation, method of transplanting and seed treatment were found to have the mean percentage score 82, 79, 74, 67, 65, 62 and 55 respectively. Practices applicable after sowing like irrigation management, post harvest technology and marketing, appropriate time of harvesting, recommended manure and fertilizer and intercultural operations secured mean percentage score of 74, 73, 71, 66 and 60 respectively.

NBFs possessed high level of knowledge about soil and field preparation of henna cultivation with the mean percentage score of

67. This was followed with mean percentage score 63, 56, 56, 63, 53 and 52 respectively for practices applied for irrigation management, high yielding variety, recommended seed rate, post harvesting and marketing, time of transplanting and recommended quantity of manure and fertilizer. For NBFs practices like appropriate time of harvesting, nursery preparation, method of transplanting, plant to plant distance, intercultural operation and seed treatment were quite different from those of demonstration farmers. Their (NBFs) mean percentage score was 50, 47, 41, 40 and 23. Regarding plant protection measures, NBFs had poor knowledge with the score of 11.

Table 8.10. Knowledge score percentage of improved agriculture practices of henna cultivation among beneficiary farmers and non-beneficiary farmers of KVK

S.No.	Improved agriculture practices of henna	Beneficiary farmers (N=60) Knowledge score	Non-beneficiary farmers (N=60) Knowledge score
1.	High yielding varieties	95	56
2.	Field preparation	82	67
3.	Seed rate	79	56
4.	Plant to plant distance	74	41
5.	Time of transplanting	67	53
6.	Nursery preparation	65	49
7.	Method of transplanting	62	42
8.	Seed treatment	55	23
9.	Interculture operation	60	40
10.	Fertilizer	66	52
11.	Plant protection	54	11
12.	Irrigation management	74	63
13.	Appropriate time of harvesting	71	50
14.	Post harvest technology	73	63

Composite fish culture: Ramasubramanian *et al*. (2011) collected data from 50 respondents from four villages in Thiruvarur district in

Tamil Nadu on the impact of KVK on the composite fish culture. The study reported a tremendous knowledge gain among the respondents. The questions for which majority of respondents wrongly answered during pre-intervention assessment were correctly answered in post intervention assessment. Ratio of soil for fish pond (from 28 per cent to 88 per cent), PH of water in fish pond (from 40 per cent to 80 per cent), nativity of carp varieties (72 per cent to 84 per cent), and seasoning of fingerlings during transit (20 per cent to 100 per cent) were the items for which the knowledge gain of respondents from interventions were found to be very higher side.

100 per cent of the respondents answered correctly to six knowledge items namely shape of ponds, season for excavation of pond, feeding habit of *mirgal,* seasoning of fingerlings during transit, acclimatization in pond and optimum time of harvest which was striking impact of intervention.

The mean score was 6.04 before intervention which got a quantum jump to 12 after the intervention. The 't' value was 11.17 significant at 1% level of probability.

Table 8.11. Distribution of respondents pertaining to their knowledge on composite fish culture before and after KVK's intervention

S.No.	Questions asked the respondents for knowledge test	Pre-intervention knowledge (%)	Post intervention Knowledge (%)
1.	Ratio of clay, alluvial and sand in fish pond	28	88
2.	PH of the water in fish pond	32	76
3.	The shape of the fish pond	80	100
4.	The number of nylon net used for inlet canal	40	80
5.	The season for excavation of fish pond	60	100
6.	The country from which Rogu got introduced to India	16	72
7.	The carp variety introduced from Japan	24	84
8.	The feeding habit of Mirgal	64	100
9.	The carp variety with fringed lower lip	52	76
10.	The carp variety with upturned mouth parts	48	88

S.No.	Questions asked the respondents for knowledge test	Pre-intervention knowledge (%)	Post intervention Knowledge (%)
11.	Seasoning of fingerlings during transit	20	100
12.	The acclamatisation of fingerlings with the rearing pond	80	100
13.	Stocking density	56	96
14.	The optimum time for harvest of fish	88	100
15.	Nature of feeding during rainy days	52	92

Overall Knowledge of Package of Improved Agricultural Practices among Farmers

The data of the studies about overall knowledge of improved agriculture practices of crops of the BFs and NBFs has been presented in Table 8.12. A cursory look of the table shows that the knowledge of the BFs about improved agriculture practices of each crop was much higher than that of NBFs. Five studies out of seven studies reported the knowledge of improved agriculture practices of crops of the BFs was from 60 to 78 per cent while only one study reported the knowledge percentage 51 per cent of paddy crop. In contrast, five studies reported the knowledge of improved agriculture practices crops of NBFs was less than fifty percent and two studies reported the knowledge of about 60 per cent. The scientists of the studies applied statistical method to find out how far there was difference in the knowledge of these two categories of farmers. They found that the knowledge of the BFs was significantly was higher than the NBFs.

Table 8.12. Overall knowledge (%) of package of improved agricultural practices among beneficiary farmer and non beneficiary farmer of KVK

S. No	Author	Crop	Beneficiary farmers (BFs)	Non-beneficiary farmers (NBFs)	Statstical test (T test/ Z)
1.	Singh (1994)	Paddy	51	27	t **
2.	Meghwal (1999)	Pearl millet	71	33	Z *
3.	Singh (1996)	Mustard	71	46	Z *

S. No	Author	Crop	Beneficiary farmers (BFs)	Non-beneficiary farmers (NBFs)	Statstical test (T test/ Z)
4.	Asiwal et al. (2008)	Mustard	78	61.	Z *
5.	Lakhera and Sharma (2003)	Mustard	61	40	Z **
6.	Sahu (2005)	Wheat	77	64.	
7.	Meena and Singh (2010)	Henna	70	47	

* Statically significant

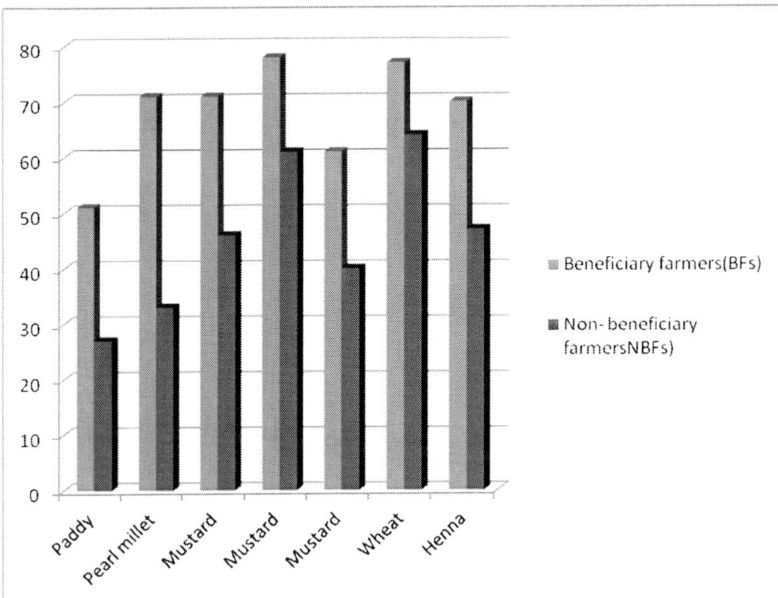

Fig. 8.4. Overall Knowledge (%) of Package of Improved Agricultural Practices among Beneficiary Farmer and Non beneficiary Farmer of KVK

Level of Knowledge of Package of Improved Agricultural practices among Farmers

Level of knowledge of farmers was classified into three groups: (i) High, (ii) Medium and (iii) Low.

It can be seen from the Table 8.13 that 62 to 70 per cent or two-third of the beneficiary farmers had medium level of knowledge and 14 to 36 per cent had high level of knowledge of improved agricultural practices of crops. The percentage of the beneficiary farmers having low level of knowledge of crops was: castor 0, mustard 2, pearl millet 2, wheat 18 and mustard 18.

In contrast, the percentage of farmers having low level of knowledge of crops among the non-beneficiary farmers was high: pearl millet 58, mustard 22 and 18, wheat 19 and castor 14. A very small percentage of the non-beneficiary farmers had high level of knowledge of crops: pearl millet 5, mustard 2, castor 9, wheat 11 and mustard 24. Most of the non-beneficiary farmers (58 to 77 per cent) had medium level of knowledge.

Table 8.13. Level of knowledge of package of improved agricultural practices among beneficiary farmers and non-beneficiary Farmers of KVK

Sl. No	Author	Crop	Beneficiary farmers (BFs)			Non-beneficiary farmers(NBFs)		
			High	Medium	Low	High	Medium	Low
1	Sahu (2005)	Wheat	14	68	18	11	70	19
2	Kumawat (2008)	Castor	30	70	0	9	77	14
3	Lakhera and Sharm (2003)	Mustard	31	67	2	2	76	22
4	Meghwal (1999)	Pearl millet	36	62	2	5	37	58
5	Asiwal and Khan (2008)	Mustard	18	64	18	24	58	18
6	More *et al.* (2002)	Cotton	28	62	10	N.A	N.A	N.A
7	Sunil *et al.* (2007)	Vermi compost	25	53	22	18	39	43

N.A. = Not available

ADOPTION OF PACKAGE OF IMPROVED AGRICULTURE PRACTICES AMONG FARMERS

The researchers asked the respondents about the adoption of individual improved agriculture practices that include field preparation, improved variety, seed treatment, sowing time, seed rate, manure, chemical fertilizer, weeding, plant protection and harvesting and storage. The adoption of individual components of improved agriculture practices of crop are discussed below.

Adoption of Individual Component of package of Improved Agriculture Practices

Paddy : Singh (1994) reported that adoption of high yielding variety (56.50%), cultural practices (60%) and irrigation (56%) by the BFs was much higher than the adoption by the NBFs. Their score was : for high yielding varieties (37%), for cultural practices (36%) and for irrigation method (36%). The adoption of chemical fertilizer by the BFs (77%) visa-a-visa NBFs (67%) was almost the same. The adoption of plant protection measure was quite low among the BFs (15%) and the NBFs (13%). This may be due to several reasons like lack of knowledge, high cost of inputs and non-availability of good quality of chemicals.

Table 8.14. Adoption score percentage of improved agriculture practices of paddy cultivation among beneficiary farmer and non- beneficiary farmer of KVK

Sl.No.	Improved agriculture practices of paddy	Beneficiary farmers (BFs) Adoption score percentage	Non-beneficiary farmers (NBFS) Adoption score percentage
1	High yielding variety	56	37
2	Cultural practices	60	36
3	Fertilizer use	73	67
4	Irrigation	56	36
5	Plant protection measure	15	13

Mustard: A study by Singh (1996) reported that adoption of ten improved agriculture practices for mustard was quite different in each of the three practices – (i) fertilizer application, (ii) time of sowing and seed rate, and (iii) recommended spacing between BFs and NBFs. The adoption of these practices by the BFs was 73, 82 and 87 per cent respectively while it was 35, 47 and 54 per cent in case of NBFs. The adoption of high yielding variety by BFs was 85 and 62 by the NBFs. The farmers adopted high yielding variety because of their conviction of its higher production potential. A moderate level of difference of adoption was found in soil and soil treatment (62 versus 42), seed treatment (62 versus 33%), irrigation management (56 versus 33%) and plant protection (67 versus 44%) between two group of farmers. Practices like weed management (42 versus 25%) and harvesting and storage (42 versus 14%) were less adopted by both farmers. Such results were found on account of poor knowledge level of the farmers because a practice like chemical weed control needs skill in its application. Since mustard seeds are resistant to storage insect pests they do not require specific control measures

Table 8.15. Adoption score percentage of improved agriculture practices of mustard cultivation among beneficiary farmer and non-beneficiary farmers of KVK

Sl.No.	Improved agriculture practices of mustard	Beneficiary farmers (BFS) Adoption score percentage	Non- beneficiary farmers (NBFS) Adoption score percentage
1.	High yielding varieties	85	62
2.	Soil and soil treatment	62	42
3.	Seed treatment	62	33
4.	Time of sowing	82	47
5.	Seed rate and spacing	87	54
6.	Fertilizer application	73	35
7.	Weed management	42	25
8.	Irrigation management	56	33
9.	Plant protection measure	67	44
10.	Harvesting and storage	42	14

Castor: Study regarding the adoption of improved practices for castor growing by BFs indicated that they scored highest for adoption of hybrid seed (79%) followed by seed rate (77%), weed management (72%) and harvesting and threshing (71).The practices – plant protection measure, use of manure and fertilizer and seed treatment – were not followed by them satisfactorily and their score was 39, 45 and 47 per cent respectively. The low adoption score these practices were due to least knowledge about the practices.

In case of NBFs the highest adoption score was in seed rate, spacing and sowing depth with 61 per cent adoption followed with harvesting and storage (58%) and weed management (52%). The practices like plant protection measure (20%), use of manure and fertilizer (24%), seed treatment (24%), hybrid seed (31%), irrigation method (35%) and sowing time (40%) were least adopted.

Table 8.16. Adoption score percentage of improved agriculture practices of castor cultivation among beneficiary farmers and non-beneficiary farmers of KVK

Sl.No.	Improved agriculture practices of castor	Beneficiary farmers (BFs) Adoption score percentage	Non-beneficiary farmers (NBFs) Adoption score percentage
1.	High yielding varieties	79	31
2.	Field preparation	69	45
3.	Sowing time	62	40
4.	Seed treatment	47	24
5.	Seed rate, spacing and sowing	77	61
6.	Manure and fertilizer	45	24
7.	irrigation management	58	35
8.	Weed management	72	52
9.	Plant protection measurer	39	20
10.	Harvesting	71	58

Henna : The BFs had maximum adoption (83%) of high yielding varieties. The mean percentage adoption score for the practices like soil and field preparation, recommended seed rate, post-harvest technology and marketing, irrigation management, appropriate time of harvesting, plant to plant distance, nursery preparation and time of transplanting was 79, 70, 69, 64 and 60 respectively. On the contrary, practices such as method of transplanting, recommended manure and fertilizer, plant protection measures, intercultural operations and seed treatment were found to be least adopted with 58, 48, 46 and 17 mean percentage, respectively.

The NBFs adopted soil and field preparation to a high extent (62%). The adoption percentage score of nine practices ranged between 41 and 54 and for plant to plant distance and intercultural operations it was 31 and 25 respectively. Plant protection measures and seed treatment were least adopted with only 8 and 6 percentage score respectively.

Table 8.17. Adoption score percentage of improved agriculture practices of henna cultivation among beneficiary farmers and non-beneficiary farmers of KVK

Sl. No.	Improved agriculture practices of henna	Beneficiary farmers (BFs) Adoption score percentage	Non-beneficiary farmers (NBFs) Adoption score percentage
1.	High yielding varieties	83	50
2.	Field preparation	79	62
3.	Seed rate	73	49
4.	Plant to plant distance	64	31
5.	Time of transplanting	60	47
6.	Nursery preparation	60	41
7.	Method of transplanting	58	44
8.	Seed treatment	17	6
9.	Interculture operation	46	25
10.	Fertilizer	58	48
11.	Plant protection	48	8
12	Irrigation management	69	54
13	Appropriate time of harvesting	68	46
14	Post harvest technology	70	43

Composite fish culture : Ramasubramanian *et al.* (2011) studied the impact of KVK in Thiruvarur district of Tamil Nadu on composite fish culture. Two dimensions of impact namely macro and micro level impact were studied. The macro level impact was assessed by analyzing the district level on three indicators namely (i) production of composite fish culture over years, (ii) number of new fish growers added over years and (iii) number of farm ponds sanctioned by Department of Agricultural Engineering over years. Carp fish production from 2007 to 2010 followed an increasing trend with at least 1000 acres increase every year. Due to focused efforts of KVK, the increase from 2009 to 2010 was 2500 acres which was 2.5 times more than the preceding year. The number of farmers who took to composite carp culture showed a significantly increasing trend with the increase of 400-500 farmers every year.

Micro level was studied under four dimensions namely (i) economic, (ii) ecological and (iii) sociological impact. Benefit : cost ratio was 1:4.25 due to adoption of composite carp culture.

There was ecological impact of the intervention of KVK. Vast majority of the respondents endorsed the statements that farm ponds to some extent used for drinking excess water (90%) and fish ponds recharge aquifers (96%). Three fourth of the respondents (70%) were aware of the disposal of farm waste as feed for fish. Majority of the respondents (88%) endorsed that effective utilization of space on the bunds of fish pond could maintain ecobalance. 36 per cent of the respondents were aware of reuse of water for growing crops.

There was also sociological impact of composite carp culture. All respondents believed that they could get to know many new techniques about fisheries as they interact with co-participants. Majority also opined that their social status in the society and among their relatives elevated due to either beneficiary of front line demonstration on composite carp culture or participants of training.

Overall Adoption of Package of Improved Agricultural Practices

Adoption measured in the context of improved agriculture practices for crops and henna. The data in the Table 8.18 show that the adoption of improved agriculture practices of crop and henna by the BFs is also significantly higher than that of the NBFs. The adoption of improved agriculture practices by the BFs as compared to the NBFs was : paddy 52 versus 34, pearl millet 59 versus 26, mustard 67 versus 39, 49 versus 33, groundnut 28 versus 23, wheat 67 versus 57, and henna 64 versus 40.

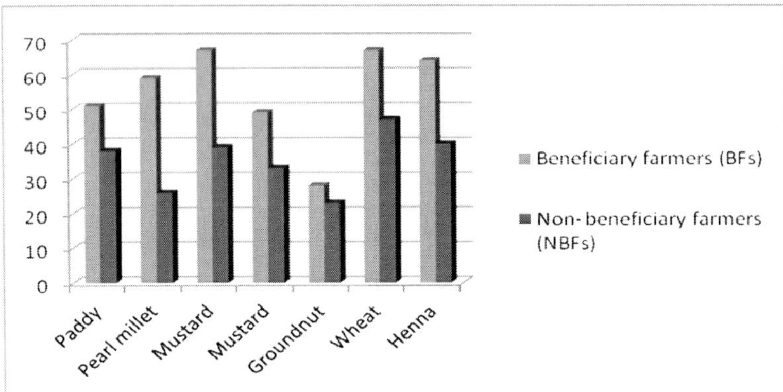

Fig. 8.5. Overall Adoption (%) of Package of Improved Agricultural Practices among Beneficiary Farmers and Non beneficiary Farmers of KVK

Table 8.18. Overall Adoption (score percentage) of package of improved agricultural practices among beneficiary farmers and non beneficiary farmers of KVK

S.N.	Author	Crop	Beneficiary farmers (BFs)	Non-beneficiary farmers (NBFs)	Statistical test (t/ and Z)
1.	Singh (1994)	Paddy	51	38	T *
2.	Meghwal (1999)	Pearl millet	59	26	Z *
3.	Singh (1996)	Mustard	67	39	Z *
4..	Lakhera and Sharma (2003)	Mustard	49	33	Z *
5.	Jondhale *et al.* (2002)	Groundnut	28	23	Z *
6.	Sahu (2005)	Wheat	67	47	-
7.	Meena and Singh (2010)	Henna	64	40	-

*Stastically Significant

The adoption of improved agricultural practices by the BFs shows that 59 to 67 percent in case of pearl millet, wheat, mustard and henna were realized, half of the improved agriculture practices of paddy, that only one fourth of improved agriculture practices of groundnut were realized,

In case of the NBFs only one fourth of improved agricultural practices of groundnut and pearl millet were realized; one third for paddy crop, mustard and henna were realized and for wheat half of improved agriculture practices were realized.

Level of Adoption of Package of Improved Agriculture Practices

Level of improved agriculture practices in Table 8.19 shows that the percent of the beneficiary farmers having medium level of knowledge of maize, mustard, pearl millet and wheat was 78, 59, 57 and 50 respectively. A significant percentage (34 to 45 per cent) of the beneficiary farmers had high level of adoption of wheat, mustard and pearl millet and vermicompost. There was no low level of adoption of improved agriculture practices for pearl millet and very small percent (7 to 16 per cent) for mustard, maize, wheat and mustard.

Level of adoption of improved agriculture practices of the non-beneficiary farmers was much lower than that of the beneficiary farmers. More than half of them (57 to 58 per cent) had low level of adoption for maize and pearl millet. 34 and 24 per cent of them had low level of adoption for mustard and wheat respectively. None had high adoption for the pearl millet crop, 3 for maize and 22 per cent for wheat. The per cent of having medium level adoption of improved agriculture practices for maize, pearl millet, wheat, and mustard was 40, 42, 57 and 61 per cent respectively.

The findings in the preceding sections indicate that the knowledge of scientific agriculture and methods of crop cultivation and their adoption among the beneficiary farmers was much higher than among the non-beneficiary farmers because of extension efforts of KVKs.

The researchers have studied the knowledge and adoption of agriculture practices of crops among the farmers. There should also be studies on the impact of improved agriculture practices of animal husbandry, horticulture, fisheries etc.

Table 8.19. Level of adoption score percentage of improved agricultural practices among beneficiary farmers and non beneficiary farmers of KVK (Percentage)

S.N.	Author	Crop	Beneficiary farmers (BFs)			Non-beneficiary farmers (NBFs)		
			High	Medium	Low	High	Medium	Low
1	Sahu (2005)	Wheat	34	50	16	22	54	24
2	Lakhera and Sharma (2003)	Mustard	34	59	7	5	61	34
3	Meghwal (1999)	Pearl millet	43.	57	0	0	42	58
4	Sharma (2002)	Maize	9	78	13	3	40.	57
5	Sunil *et al.* (2007)	Vermi compost	45.00	40.00	15	*	*	*

There is no study on skill development of farmers in agriculture technology. If the farmers can not use the knowledge gained due to lack of skill, they may not be interested in any extension programme.

The KVK in Uttar Pradesh modified the old iron – made mould board plough by reducing its weight, size and thus the price. This plough fitted the condition of the draft power of the available small animals (oxen) on one hand and substantial reduction of price of the plough from Rs. 360 to almost Rs. 40 on the other. The plough became popular very popular - they were not adopted earlier with the exceptions of the big farmers who met both conditions; (1) better animal power; and (2) Money. A **Subsoil water injector** invented by a farmer in Tamil Nadu and later modified by the scientists is popular equipment for farmers for farmers for dryland horticulture and plantation crops. it worked on the same principles of medicine injection in human body. A liter of water is right-way injected in the root zones of the plants without any loss to occur by evaporation. The cost of subsoil water injection is about Rs 600 (Prasad and Reddy, 1991).

TRANSFER OF AGRICULTURE TECHNOLOGY FROM THE BENEFICIARY FARMERS TO NON-BENEFICIARYFARMERS

It is expected that the improved technologies taught in front line demonstration to the participating farmer to educate their neighbors. By this trickled down effect maximum number of no-participants are also expected to learn and follow improved agriculture technologies. The study by **Kushwaha and Indra Pal** (1998) in KVK Fatehpur in Rajasthan revealed that the farmers participating in front line demonstration had not only gained knowledge through demonstration, but on improved technologies oil seed cultivation also helped the non-participating farmers in their neighborhood to gain knowledge.

AGRICULTURE PRODUCTIVITY

Rai ***et al.*** (1999) studied the role of front line demonstration (FLD) in transfer of improved agriculture practices of linseed cultivation conducted by KVK Chhindwara in Madhya Pradesh. The FLD were conducted at 55 farmers' fields during the period from 1990-91 to 1994-95. The yield of linseed in FLD was significantly higher than farmers' method of linseed cultivation. The yield (13.75 to 19.40 q/ha) of FLD was 114 to 169 per cent more than local farmers' practice yield (5.25 to 7.5 q/ha), 277 to 486 per cent district yield and 409 to 532 per cent state yield.

Table 8.20. Yield performance of linseed crop in demonstration.

Year	Yield (q/ha)				% increase in yield over		
	Demonstration	Local farm	District	State	Local	District	State
1990-91	13.75	6.40	3.65	4.09	115	277	409
1991-92	16.46	6.90	2.81	4.56	138	486	456
1992-93	15.60	6.20	3.56	4.26	152	338	426
1993-94	19.40	5.25	4.17	5.32	169	365	532
1994-95	17.95	7.05	3.47	4.54	154	417	454

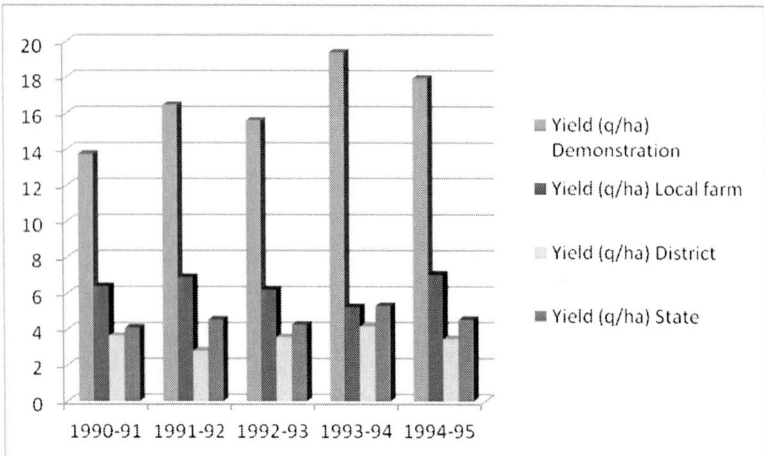

Fig. 8.6. Yield performance of linseed crop in demonstration. (*Source:* Rao, *et, al.,* Role of front line demonstration in transfer of linseed technology, Maharashtra Journal of Extension Education XVIII, pp. 277-278, 1999).

Rao and Chandrasekhara (2010) studied productivity enhancement in groundnut through FLD in KVK Visakhapatnam. The KVK conducted FLDs with improved varieties along with recommended varieties. Two varieties used were 'TAG 24' and TCGS 320' were released at national level and state level, respectively. FLDs were conducted at Yellamanchili Mandal from 2003 to 2005 for variety TCGS 320 of groundnut in an region of 29.80 ha in 82 locations and Devarapalli Mandal from 2004 to 2007 for variety TAG 24 in an area of 30 ha at 86 locations.

The study reported that the varieties TCGS 320 and TAG 24 were superior to local variety. TCGS 320 with improved package practices gave 22.80% to 44.39% increase in yield over local variety with an average improvement of 35.03% and 5.67% to 32.20% increase in yield over district average with average improvement of 24.57.

In case of 'TAG24' the yield improvement ranges from 15.38% to 47.98% with an average improvement of 31.33% compared to local variety and yield improvement ranges from 15.20% to 36.55% compared to the district average with an average improvement of 25.86%.

Table 8.21. Yield performance of linseed crop in demonstration

Variety / Season	Yield (q/ha)			% increase in yield over	
	Demons-tration	Control	District	Control	District average
TCGS 320 Kharif	14.93	10.84	10.43	27.39	30.14
Rabi	17.80	14.50	14.00	22.80	21.34
Kharif	15.79	11.13	12.29	41.86	28.47
Rabi	18.25	13.23	17.27	37.94	5.67
Kharif	14.45	10.64	10.93	35.80	32.20
Rabi	16.56	12.15	13.15	35.03	24.57
Average	16.56	12.15	13.15	36.03	24.57
TAG 24 Rabi	27.22	17.33	1727	36.33	36.55
Kharis	15.21	12.42	10.93	22.46	28.13
Rabi	16.51	14.35	14.00	15.38	15.20
Kharif	13.56	10.87	10.57	24.74	28.28
Rabi	17.64	11.92	15.26	47.98	15.59
Kharif	14.72	10.43	11.20	41.13	25.86
Average	17.48	12.89	13.20	31.33	25.86

Source: Rao and Chadrasekhara, Productivity enhancement in groundnut through Front Line Demonstration, Indian Farming, 60, 4, pp. 19-21, 2010.

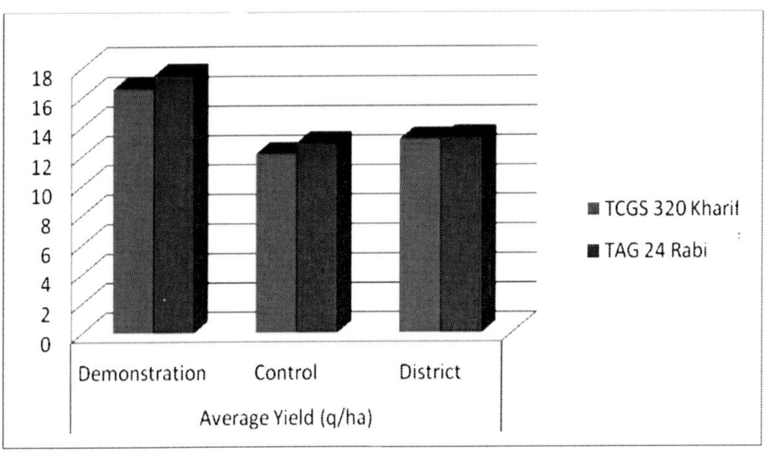

Fig. 8.7. Yield performance of linseed crop in demonstration.

Das (2002) studied effectiveness of Self-Help Group (SHG), a method of extension taken by KVK in Gadag district of Karnataka. The procedure followed by organizing the SHGs were– (1) Consultation meetings, (2) Training and exposure visits, (3) Conducting awareness meetings, (4) Identification and recognition of group, (5) Stabilization and sustenance of the group, (6) Follow up and capacity building, (7) Use of groups for training, front line demonstration, organizing extension activities, production of seeds and planting material and (8) Linkage with financial and other institutions. The following table shows effectiveness of SHG mode of operation over the individual farmer approach in respect of the following interventions.

1. Balance nutrient management in groundnut;
2. Milk production on buffalo rearing by enrichment of dry fodder locally available;
3. Value addition in chillies;
4. Addition occupation of buffalo rearing for SHG.

Table 8.22. Effectiveness of group approach (Self Help Group) on agriculture productivity

S.N.	Practices	Yield (q/ha)	Net income (Rs/ha)
A.	**Adoption of balanced nutrient management in groundnut by self-help group**		
1.	Before-imbalanced and untimely	6.25	6875
2.	After-Balanced and timely (NPK, gypsum and micro-nutrient)	1175	9456
B.	**Increase In milk production of SHGs on buffalo rearing**	**Milk yield (liters/day/ buffalo)**	**Net returns (Rs/day)**
1.	Existing practice in non-members	2.00	18.00
2.	Practices followed after interventions through SHGs	4.00	32.00
C.	**Value addition in chill-based cropping system**	**Dropped chillies/ha**	**Profit/ha (Rs)**
1.	Existing	75	250
2.	Value addition by SHGs (masala chili)	75 kg converted into 20 kg	350

D.	Increase in income of SHGS due to additional occupation of buffalo rearing	Income/day (Rs)	Additional income (Rs)
1.	Before- stone cutting and wage work	20.00	-
2.	After- Stone cutting, wage work and buffalo rearing	48	900.00

Source: Das, Effectiveness of group approach and information technology in micro-level dissemination of agriculture technology, Indian Farming, 52, 8, pp.26-32, 2002.

DEVELOPMENT OF ENTREPRENEURIALACTIVITIES

Bharti and Dhriti (2006) made study on the role of KVK in promotion of entrepreneurial activities among rural women in ten villages of Hanumangarh district of Rajasthan. 16 enterprises had been promoted in the villages. Nine enterprises which were exclusively run by women were taken for the study. Total number of trained women was 150 on these enterprises. All these were taken for the study. 100 per cent women adopted *surf, Masala* and *raakhi* making enterprises. 88 per cent women adopted stitching enterprise while 97 per cent women adopted tomato preservations enterprise. 64 per cent women adopted *dari* and *knees* making enterprise, 66 per cent women adopted herbal beauty parlor and candle making. Toys making enterprise was adopted by 43 per cent women.

Some of the women did not adopt an enterprise or discontinued after adopting it. In case of *dari* and *knees* making, low return from it was the most important reason for non- adoption. With regard to herbal beauty parlor, soft toys making, candle making and stitching, lack of time was the main reason for non adoption. A few women mentioned that due to study they were unable to spare time for the activity.

When some of them discontinued, as was in case of the women who left the *dari* and *knees* making and soft toys making it was due to less demand in the market. In case of herbal beauty parlor there were three women who left the enterprise due to difficulty in obtaining the material (cosmetic) required for the parlour. They reported that the required cosmetics were not available in the village and they had to go to the city for buying the required cosmetics. The women who left the candle making enterprise reported lack of time as the main reason for discontinuance. In case of stitching there was only one woman who discontinued the enterprise due to her marriage.

Majority of the women entrepreneurs were satisfied with the role played by the KVK in providing technical guidance, organizing the training, financial assistance, linkage with other agencies, marketing of produce and follow-up activities.

CASE STUDY

A Karnataka farmer, Shri Umesh, hailing from Kalya village of Magadi taluk of Ramangara District, has become a model for the impoverished farmers of the state. A small farmer with 2.50 acres, earned a net annual income of Rs 4.80 lakh. Initially Shri Umesh's traditional farming included only arecanut in one acre along with a small dairy unit. The technical support from KVK Ramangara helped him to shore up and change his farming through introduction of improved crop varieties, stall feeding of sheep, upgrading dairy unit, growing azolla as cattle feed, vermicompost manufacturing, poultry, developing a fodder bank and drudgery reduction through farm mechanization at his level. The most crucial intervention was the introduction of rose and marigold cultivation along with vegetables like pole beans, cucumber, tomato and brinjal as intercrop. He has a stall feeding of 17 sheep that has drawn the attention of villagers.

During the last three seasons Shri Umesh earned Rs 2,40,000 from vegetables alone. Everyday he sells flowers in the nearby market earning Rs 800-1,000 per day.

Keeping the requirement of the market demand he planned two varieties in such a way that they start flowering in September – October when the demand for fresh flowers would be high in the state. Earlier, the only source of income for him was from arecanut which fetched him a net income of Rs 1,45,000 annually.

He was advised to establish a fodder bank comprising CO-3, CO-4 grass varieties and Azolla to feed his small dairy unit (one HF and one Jersey cow) and sheep unit. He has also planted silver oak trees all along the borders of his farm. The trees act as a wind barrier to the arecanut garden. KVK guided him in preparing his own feed mixture for the sheep as well as dairy animals. The combination of azolla and the feed mixture has reduced his feed cost by Rs 150 per day. The entire garden is irrigated through a micro sprinkler. To reduce drudgery, the farmer has opted for mechanization by deploying a chaff cutter, rotowater and a cycle weeder that has helped him to reduce the cost of labour.

He recycles farm waste into vermi-compost. The average production from his farm per year is 0.8 ton of green areca nuts about one tone of vermi-compost, 15 tonnes of cow dung, 40 tonnes of fodder grass and vegetables worth Rs 2,40,000.

Till about three years back he did not even own a two wheeler, but today from the income generated at his farm he owns a car. His successful farming has already drawn the attention of hundreds of farmers within and outside the district (Prabhu, 2014).

Another farmer, S. Thinakarasamy from Kooliyankadu village of Pundukkottai district, owns three acres of wetland and three acres of dry land. With a strong desire for learning the latest techniques in nursery production, he attended several training programmes run by KVK Vamban, Pudukotai on propagation of ornamental plants, like grafting casuarinas, eucalyptus, grafting of jack fruit and cashew and other fruit crops and started a nursery on a small scale. He was advised to construct a polyhouse at his farm and start propagation of ornamental plants like duranta, crotons, jasmine and medicinal plants like coleus and supply them to Pudukottai town. He has succeeded in doing these. Meanwhile, he was approached by farmers of nearby villages for guidance on growing fruit and forest plants. He is a trusted and famous nursery man in Pudukottai district for sourcing seedlings. He also carries out sowing seedlings in the farmer's field on contract basis.

His socio-economic status improved after he entered into the nursery venture. Since he started the nursery in 2007 he has sold more than ten lakhs seedlings and earned a net profit of Rs 7,14,750 every year. He bought a new two wheeler and later a three–wheel vehicle for nursery and transporting seedlings. At the time of study he was engaged in establishing a separate nursery for medicinal plants.

While S. Thinakarasamy's quantum leap from subsistence to well to-do farming, another farmer, B. Thamilsekaran also started a small nursery unit for propagation of jack grafts. Now he is a famous nursery man in his village and is in demand for supplying jack grafts.

His earning is around Rs 500 per day. His annual income from jack graft rose to Rs. 3.51 lakhs which was 50 percent more than his income from other sources (Prabu 2014).

Appal Narshamma, a resident of village Manglutan in Andaman district, constructed a pond (35 m × 15 m × 3 m) to irrigate her plantation, use water for poultry birds and also fish culture. She did not arrange to get quality fish seed in her pond. Then she approached the KVK of Central Agricultural Research Institute, Port Blair and expressed her interest in freshwater fish culture. She discussed with them her past difficulties in fish farming. The scientists of the Institute visited the pond and it was found that the pond was infested with tilapia fish. Then, it being the monsoon season, it was not possible to de-water the pond. Therefore, by repeated netting most of the fish were removed. The pond soil was treated with required lime and for an initial trial only 150 fish fingerlings (25 catla, 50 rohu and 75 mrigal) were stocked in the pond. She was advised to follow scientific culture practices regularly. In the mean time, she also attended the training in fish culture organized by the KVK. Other critical inputs like feed and fertilizer were supplied by the Institute free of cost as incentives for the first attempt. After seven months of culture the pond was harvested in March 1998 and she could get carps of a average 230 g and 10 kg of tilapia. The carps were sold at the rate of Rs 45 /kg. This production and further interaction with KVK workers prompted her to go for higher stocking and scientific farming for better fish production. In 1997, she de-watered the pond completely and stocked the pond with quality fish fingerlings. She followed the recommended techniques and could harvest 85 kg of carps. This crop fetched her Rs 4,675 @ Rs 55/ kg with a total expenditure of Rs 1,450 toward the cost of fish seed, feed, lime and fertilizer. The net income was Rs 3,225 (Raj *et al*., 1999).

The KVK of Central Agricultural Research Institute, Port Blair surveyed the villages of South Andaman to asses the training need of the farmers and farm women. In one of such interaction Malathi Roy, wife of Shri Sukhranjan Roy of South Andaman expressed her desire to undergo training imparted by the KVK on 'nutritious kitchen gardening' . Utilizing the opportunities, she attended several training programmes on different vegetable cultivation methods, including those on application of chemical fertilizers and on control of insect pests and diseases. In Andaman, climatic factors are very congenial for the rapid multiplication of the insect pest diseases. A few of the major insects that attack vegetables are fruit borers, leaf caterpillars, mealy bugs, aphids and the diseases are bacterial blight, root –rot, etc. Malathi Roy learnt the ways of controlling the above insect pests and disease. She also consulted the field experts from time to time.

She cultivated vegetables in an area of 43 m x 41 m. She got Rs 15,000 from amaranths in December, after which she raised okra on the same land and got Rs 14,000 in February. Along with these vegetables, she grew bottle gourd also and earned Rs 10,000 in March –April from the same land. Thus she earned Rs 40,000 from vegetables within a short period, whereas her previous income (before the adoption of KVK suggestions) was only Rs 5,000 from growing a single crop. The enhanced income brought her appreciable respect from her family members who recognized her role as an high income –earning member of the family (Raj *et al.*, 1998).

SECTION 3
ATTITUDE OF THE PERSONNEL Of KVK TOWARDS KVK

Saini and Pandia (1997) studied to know the attitude of the training staff of Krishi Vigyan Kendras towards its functioning. The respondents were the Chief Training Organizers and training Assistants of 89 KVKs in the country. They found that 88 per cent respondents had a favorable attitude toward towards philosophy, concept and principle of KVK. About 68 to 75 percent respondents had favorable attitude toward the objective of KVK, linkage with other organization and training methods used in the transfer of technology. Eighty per cent respondents had neutral attitude towards their fellow staff members. However, 10 per cent had unfavorable attitude toward their fellow staff members. 49 per cent respondents had neutral attitude toward the training facilities, 15 per cent had unfavorable attitude towards the training facilities, and only 37 per cent respondents had favorable attitude toward the training facilities. This shows that training facilities at the KVKs were lacking and should be strengthened.

Kumar and Singh (1994) conducted a study in three states – Bihar, West Bengal and Uttar Pradesh – and collected data from 100 trainers from KVKs through questionnaire. They found that the trainers had moderate level of satisfaction regarding coordination with agriculture university and non–officials. They were slightly satisfied with coordination with state development agencies, the method of fixing target and competence of superiors. The items showing least satisfaction were timeliness of supplies, development policies related to professional work, quality of supplies, professional work and the time taken by superiors to solve professional problem. The trainers expressed dissatisfaction with time taken by authorities to solve personal problems. Thus there is need to plan ways to enhance quality of the precious human resource manning the KVK, with inputs such as rewards, incentive and motivating work environment.

Singh and Prasad (1997) carried out a study on quality of working life in eight KVKs in Bihar in which four KVKs were managed by voluntary organizations and the remaining four by Agricultural Universities. The respondents included 40 scientists and 24 technical cadre personnel. The KVKs in Bihar gave a score of moderate level of satisfaction in quality of working as the mean score was 85.50 out of the maximum value obtainable score of 125. They suggested that the working conditions in the KVKs need to be improved. The quality of working life of personnel of KVKs run by voluntary organizations was significantly superior to those of KVKs run by agricultural universities. It warrants that authorities of agricultural universities should study the reasons and take steps for improving the quality of working life of the KVKs organization. On the whole, the study suggested the need for looking into entire aspect of KVKs management to raise the level of quality of working life of the personnel. Achievement motivation and job involvement emerged as potent factors affecting the quality of working life.

ATTITUDE OF THE FARMERS TOWARDS KVK

Kumar *et al.* (2006) made a study of the attitude of farmers towards the activities of KVKs in five villages in Sirohi district of Rajasthan. The total number of respondents who were beneficiaries of the programmes was 50. 32 per cent respondents had most favorable attitude toward the various activities of the KVKs, 48 per cent had favorable attitude and only 20 per cent had least favorable attitude.

Gaikwad and Gunjal (2000) in their study reported that all the beneficiaries appreciated the activities of the KVKs, but also stated that more incentives were provided by NGO KVKs than SAU KVKs. Management orientation and information seeking behavior among the beneficiaries of NGO KVK was found better than SAU KVKs. All the KVKs conducted sufficient technical programmes in all the disciplines, however, NGO KVKs gave emphasis to diversified agriculture activities like horticulture, animal husbandry, along with the knowledge of home science.

The study of **Chauhan *et al*.** (2007) based on 45 trainees in four KVKS in Rajasthan revealed the following major constraints : (i)lack of publicity of training programme, (ii) non-availability of required input, (iii) lack of coordination with other agencies, (iv) lack of need assessment and (v) lack of transport facility for field visits during the training programme.

Jonadhale and Sangle (1996) reported on the problems and constraints in KVK programmes. At the top was problem of the transportation mentioned by 75%, routine work affected by attending the training programme (64%) and the amount of stipened was insufficient (50%).

SECTION 4
SUGGESTION FOR EFFECTIVENESS OF KVK

National Commission on Agriculture (1976) in its report suggested following measures for better working of KVKs.

Each of the KVKs should have academic freedom to develop its own programme in conformity with the needs of the area in which it is situated. For this purpose, it would be advisable to consult local authorities and seek their cooperation on matters concerning them. The training programmes have to be need based courses of short or long durations varying from a few weeks to a few months, or even longer. These may be part-time and full-time education course. Some of these, specially the long duration ones, may be conveniently be correspondence courses for the farmers.

Evaluation should be built-in component of the KVK programmes so as to enable the KVKs to revise and improve the courses in conformity with the need of the region, type of trainees and the nature of development programmes.

Tribal areas having a large percentage of dropouts at the primary/secondary levels of education need special attention. Some programme of non-formal vocational education suitable for the local conditions should be formulated which would in several respects different from the ordinary ones. Only those who have knowledge of problem of the tribal areas and of the people should be entrusted with these programmes.

For the success of the programmes of vocational agricultural education a few conditions need to be satisfied. Learners attached to a production enterprise should be suitably remunerated. Educational facilities created should be linked with developmental programmes and availability of job openings. As a follow-up measure, the trainees not absorbed in jobs should be provided with loans and other facilities like marketing, extension and farm input services, as a step towards self- employment. In this way, trainees would serve as innovative farmers and set examples for others.

Vocational education is comparatively costly. Therefore, infra-structure of existing institutions should be utilized, as far as practi-cable, for this purpose. For vocational courses meant for women trainees preferably women teachers should be appointed. The success of vocational education programmes is pivoted on the availability of competent and devoted teachers. But the problem of recruitment of the right type of teachers can not be solved easily. Suitable emolu-ment is no doubt an important incentive and should be provided. At present some of the teachers are being trained in the teachers' training colleges, but their necessary orientation towards agriculture is usually limited, or even absent. Moreover, the trainee teachers do not possess sufficient past experience in farming or related trades. In the circumstance, it will be advisable to give preference to those, who have graduated from polytechnics and vocational agricultural schools and have appreciable farm experience. These teachers could best be trained in polytechnics attached to the agricultural universities.

With a view to meeting the needs of development in agriculture and related activities, it was recommended that there should be one centre for each district to provide for regular short duration training in agriculture and agro-industries. Such a centre may be a KVK. If any existing agricultural education institution has the potential, it should be developed into a KVK. There should be at least one KVK in each district by 1985 and at least three for each district by 2000. If the institution of the KVKs has to serve a useful purpose in modernizing agriculture in the country and in making an impact on agriculture, the number of KVKs has to be increased.

According to **Dr. A.N. Shukla** (1991), Former Assistant Director General Extension, ICAR, New Delhi, the KVKs, though they have proved their utility and credibility among the farming community, suffer from a number of administrative, physical and financial constraints. In the absence of proper incentive mechanism, preference is not given by the scientist to extension training of farmers. This phenomenon becomes acute if the projects are located in remote rural areas. The KVK scientists working with voluntary organizations suffer and get de-motivated for want of uniform and attractive service conditions and privileges. With the increase in participation of farm-women, KVKs feel the need of having separate hostel facilities for farm-women, which will encourage this segment of rural population to participate in training and development efforts. Similarly budgetary provision and the norms of stipend to the trainees of on-campus trainings are very meager. This needs enhancement to meet their expenses on transport and boarding facilities.

The jurisdiction of a KVK is the district in which it is located. According to a rough estimate, the number of farmers, farm –women and rural youths in a district comes to about 6 to 7 lakhs, which is a very large number to be covered by one KVK. On an average, one KVK could train approximately 2,500 to 3,000 trainees effectively. According to this, one KVK may take more than two centuries to cover each and every farmer living in its jurisdiction. In view of the gigantic task, National Commission on Agriculture (1976) rightly recommended the establishment of one KVK in each district of the country by 1985 and three KVKs in a district by the end of 2000 AD. Further, a High Powered Committee on Agricultural Policies and Programmes appointed by the Government of India (1990) recommended that the immediate goal should be to establish at least one KVK in each district of the country.

Dr. C. Prasad (1991), Former Deputy Director General (Extension), ICAR, New Delhi, suggested the following measure for efficiency of KVKs.

The KVK must demonstrate all relevant technologies relevant for the area on its campus farms as well as in the neighboring villages for demonstration as well training.

It is generally believed that that the farmers are a very receptive and they are ready to accept technologies for their economic betterment. This is a blanket statement and only partially true. In this context, perhaps, we are talking of 16-20 per cent of farming population of the normal adoption curve – the innovators, the progressive, the rich; the rest, almost 80 per cent of the farmers are either adopters, slow-adopters or laggards. It is the latter group (80 per cent) which needs to be targeted by training and extension systems and not the innovators. The innovator farmers are themselves capable of not only helping themselves, but also serving as the agents of change if rightly motivated and guided. If adoption of technologies was so easy, after 40 years of organized extension services in the country, the farmers would have adopted at least low-cost and no-cost agricultural technologies like, line sowing, compost making, soil testing, crop rotations green manuring, plant population, balanced fertilization etc. in large measures. It is here that the need for training of farmers in the rudiments of scientific farming becomes apparent.

Unless the farmers are properly trained and educated, they are going to be a basic limiting factor in future in accelerating agricultural production; the improved agricultural technologies in agriculture may not be sufficient without the farmers' education.

It may be beneficial to have Home Science Unit as a separate small complex with a the lecture room, sitting place for teachers (2), home science laboratory, hostel for 20-25 girls, quarters for the two teachers, store etc. all integrated in one building with its own boundary, play ground, kitchen garden, nursery school etc. For long – duration residential courses – six months or so, this facility will go a long way in preparing our young rural girls for bright socio-economic future and better home–management.

There should also be, as a part of main building, a hall to accommodate 200-300 farmers for larger meetings, functions, seminars, etc. In all cases, the buildings have to be modest economical and very much in harmony with rural setting and design. They must not be copies of urban design. The building can also be used for exhibition and as an information centre.

Training at KVK can not be a light hearted effort: it cannot be allowed to become a routine affair. For the KVK staff every training course should be an experiment; the trained staff or extension personnel should have the farmers as their long–term clientele; and growth of farmers, farm women and young farmers should be their achievement. This cannot a short – time goal. The KVK staff cannot be satisfied merely by completing training courses. Their concern should be deeper, multi dimensional, research oriented and larger benefits to the farmers. Innovation is the outcome of creative thinking and work. To make KVKs innovative, the KVK scientists should be basically researchers in the area of training. The KVKs work should be taken as an action research.

"Teaching by doing" and **"Learning by doing"** - the design of KVKs is of great significance. It will take our best efforts to attain success in farmers' training programmes and courses. The Chief Training Organisers need to be constantly watchful to assess the success or failure of their work. Their efforts should be examined and see that the lapses and gaps in trainings are removed. KVK will need a lot of creativity in fulfillment of their mission.

The training design of KVKs includes using progressive farmers as the resource trainers. Such courses for the farm youth can be offered on the fields of the farmers as long- term courses of 3 to 6 months. The trainer farmers may be formally recognized by some token fees and stipends may be given to the farm youths who are being trained. The KVK teachers can supervise such off-campus courses. This will multiply the gains and the training in situ will be most relevant for the trainees. An experiment should be done to see how this approach works.

There should be in-built system in the organization for updating and upgrading knowledge of the KVKs staff through various training courses.

Gaikward and Gunjal (2000) made a study to find out the constraints faced by the KVK staff and beneficiaries of the KVK. The study was conducted in four KVKs – one each from four regions of Maharashtra.

Constraints faced by the KVK staff: All the KVKs reported the following constraints : (1) insufficient fund for training and meals of the participants, (2) insufficient provision for mobility, (3) no provision for maintenance of the buildings, (4) insufficient revolving funds, (5) no facility for giving loans to youths to start their enterprises. Besides these the Agricultural University run KVK had the following problems: (1) no irrigation facility at farm, (2) frequent transfer of staff, (3) the KVK staff not staying at the campus, (4) non-availability of labour and (5) delay in releasing the funds.

Constraints faced by beneficiary of KVK: 52 per cent trainees from all KVKs reported economic constrains, such as unavailability of loan, high rate of interest on loan and less incentives for agriculture and high cost of labour. More than 50 percent beneficiaries reported the following technical problems : negligent attitude of extension workers towards the farm, lack of training facility, non-availability of literature on agriculture, non-availability of inputs and service. There were social problems also, such as lack of group formation, illiteracy and lack of participation of women in getting agriculture information. More than 65 per cent beneficiary farmers reported various market problems like low rate given by the middle men, non-availability of transport and having no processing unit for agriculture produce.

The suggestions for improving the activities of KVK are presented below.

1. More than 50 percent beneficiaries suggested that the loan should be made available at a lower rate, also the technology employed, and labour should cost less thus brining down the cost of cultivation.
2. There should be more demonstration on new agricultural technology, better organization of exposure visit to show innovative techniques and also more self employment generating courses. There should be more frequent visits to farmers' fields, and improved availability of various services at local level as well as production and distribution of agricultural literature. There should be more emphasis on making various inputs and services available to the farmers so that they can easily adopt the new technology.
3. There should be formation of group of farmers so that transfer of agricultural technology should be easy. More farm women should be involved in getting agriculture information.

4. About seventy percent beneficiaries suggested that market information should be provided at local level so that they could send their agricultural produce most profitably. There should be market facility nearby their villages for them to sell their produce easily.

According to the study of **Sharma** (2002) in Raigarh in Madhya Pradesh the trained farmers wanted increase in the amount given for meal of the trainees. This was considered the foremost important measure for improving the farmers training programme. In each training programme practical knowledge should be provided through interactive demonstration. 60 per cent trained farmers reported that KVK must inform the selected trainees well in advance about duration and topic so as to help farmers to prepare themselves before the training. The other measures suggested by the trainees were that number of trainers must be increased. Trainers must be technically sound, sufficient furniture should be available at farmers' hostel as well as training hall and KVK should have a demonstration unit. Printed material at the training time and proper transport facilities should be available with KVK because many a times field visit had to be cancelled due to non-availability of vehicle. Long duration programme should be organized and number of trainees per course may be restricted up to 20-25.

According to **Singh and Singh** (1995) the following two factors should be considered in the selection of trainees for the training programme: (1) the farmers should have a reasonably good education standard, (2) the farmer should have farm power at their disposal along with good contact with extension worker.

REFERENCES

Asiwal, B.L., Singh, Sangram and Khan, I.M., "Knowledge Level of Beneficiary and Non- beneficiary Farmers of FLD Regarding Improved Mustard Production Technology in Sikar District of Rajasthan", Rajasthan Journal of Extension Education, 16, pp. 119-123, 2008.

Bari, Mohan, "KVK Scientists Generate Employment in a Tribal Village, Indian Farming", XXXVII, 7, pp. 22, 1987.

Bharti, Anshul and Solanki, Dhriti, "Role of KVK in Promoting of Entreneurrial Activities among Rural Women", Rajasthan Journal of Extension Education, 14, pp. 144-148, 2006.

Chauhan, Ram Singh and Singh, Sangram, "Constraints Faced by Beneficiaries in Krishi Vigyan Kendras of Rajasthan Agricultural University", Rajasthan Journal of Extension Education, 15, pp. 48- 50, 2007.

Chauhan, R.S., Dogra, S. and Lathwal, O.P., Impact of Training on Adoption of Farm Technology, National Seminar on Information and Communication Technology : Opportunities and Challenges for Revitalizing Extension System – December 27-29, 2006, Navsari (Gujarat), pp. 248-249.

Das, P., "Effectiveness of Group Approach and Information Technology in Micro-level Dissemination of Agricultural Technology", Indian Farming, 52, 8, pp. 26-33, 2002.

Deshmukh, Nanaji, Extension Approach for Integrated Development of Small and Marginal Farmers in Rainfed, In Extension Strategy for Minimizing Risk in Rainfed Agriculture Edited by Prasad and Das, P., Indian Society of Extension, pp. 387 -392, 1991.

Desai, B.R., Firase, K.A. and Kokate, K.D., "Appraisal of Training Programme of Krishi Vigyan Kendra, Maharashtra Journal of Extension Education", XV, pp. 44-79, 1996.

Dare/ICAR Annuual Report, 2008-09 to 2013-14.

Farooqui, H.F., Katare, P.M. and Kulkarni, M.V., "Training Needs of Farm Women", Maharashtra Journal of Extension Education, XI, pp. 257-262, 1992.

Gaikwad, B.H. and Gunjal, S.S., "Information Seeking Behavior of KVK Beneficiaries in Maharashtra", Journal of Extension Education", 11, 1, pp. 2672-2676, 2000.

Gaikwad, B.H. and Gunjal, S.S., "Constraints Faced and Suggestion Made to Improve Activities of the Krishi Vigyan Kendra in Maharashtra", Indian Farming, 49, 12, pp. 34-35, 2000.

Ghosh, P.K. and Pandey, K.N., "Impact of Training on Knowledge of Farmers about Improved Rice Cultivation Technologies", Indian Journal of Extension Education, XXXIX, 1 & 2, pp. 108-110, 2003.

India 2011, A Reference Annual, Publication Division, Ministry of information and broadcasting, Government of India, pp. 66-115, 2011.

Jonadhale, S.G., Ingale, L.A. and Fatak, U.N., "Impact of Krishi Vigyan Kendra Training on Adoption of Improved Practices of Summer Groundnut", Maharashtra Journal of Extension Education, XIX, pp. 109-111, 2000.

Jondhale, S.G. and Sangle, G.K., "Evaluation of KVK Programme and Constraints Faced in Its Execution", Maharashtra Journal of Extension Education, XV, pp.211-218, 1996.

Jondhale, S.G. and Bhele, W.L., "Training Needs of Farm Women in Krishi Vigyan Kendra", Maharashtra Journal of Extension Education", XVII, pp.294- 298, 1998.

Khan, A.K., Singh, G.P. and Choudhary, A.N., "Training Needs of Naga Farmers in Terraced Rice Cultivation in Kohima", Maharashtra Journal of Extension Education, XII, pp. 381-382, 1993.

Kumar, Mahendra and Khan, I.M., "Training Programmes of KVKs : Opinion of Farmers", Indian Journal of Extension Education, 41, 3 & 4, pp. 110-112, 2005.

Kumar, B. and Singh, D.P., "Profile of KVK Trainers", Indian Journal of Extension Education, XXX, (1 & 2), pp. 42-49, 1994.

Kumar, Deleep, Kalla, P.N. and Dangi, K.L., "A study of Attitude of Farmers towards Various Activities of KVK, Sirohi, Rajasthan", Rajasthan Journal of Extension Education, 14, pp. 106-108, 2006.

Kumar, Mahendra and Khan, I.M., "Training Performance of the Krishi Vigyan Kendras by Different Agencies in Rajasthan", Rajasthan Journal of Extension Education, 14, pp. 41-46, 2006.

Kumawat, S.R., "Impact of Front Line Demonstration on Adoption of Improved Castor Production Technology", Rajasthan Journal of Extension Education, 16, pp.143-147, 2008.

Kumbhare, N.V. and Khonde, S.R., "Impact of KVK Training on Farmers Adoption Behaviour and Knowledge Gain", Indian Journal of Extension Education, 45, 3& 4, pp. 60-62, 2009.

Kushwaha, R.K. and Pal Indra, "Knowledge level of Participants and Non-Participants of Mustard Front Line Demonstration, In National Conference on Extension Education for Sustainable Development, Mumbai, December 29-31, 1998, pp. 6.

Lakhera, J.P. and Sharma, B.M., "Effect of Front Line Demonstration on Transfer of Improved Mustard Production Technology", Rajasthan Journal of Extension Education, XI, pp. 89-92, 2003.

Mangat, Ramandeep and Gill, L.S., "Krishi Vigyan Kendra : Creating Self-Employment and Generating Income in Punjab", Kurukshetra, 47, 4, pp.35-38, 1999.

Meena, M.L. and Singh, Dhreeraj, "Impact of Training and Demonstration in Adoption of Henna Production Technology by Farmers", Journal of Extension System, 26, 1, pp. 84- 87, 2010.

Meghwal, Rewat Ram, Impact of KVK in the Adoption of Improved Practices of Bajra Cultivation In Jodhpur District, M.Sc. Thesis, pp. 1-103, 1999.

Nanda, Rakesh and Tantray, A.M., "Participation of Women in KVK Programme", Indian Journal of Extension Education, XXVI, 1&2, pp. 122-124, 1990.

Narayanswamy, C. and Eshwarappa, G., "Impact of Front Line Demonstration", Indian Journal of Extension Education", 34, 1 & 2, pp.14-15, 1998.

Prasad, C., Krishi Vigyan Kendra – A Land Mark for Agricultural and Rural Development, In Extension Strategies for Rainfed Agriculture, 1991, Edited by C, Prasad and Das, P. Indian Society of Extension Education, International Fund for Agricultural Development, Indian Council of agricultural Research pp. 371-376. 1991.

Prabu, M.J. Generating Constant Income Imperative for Successful Farming, Hindu, 4, 258,, pp.14, October 31, 2014.

Prabu, M.J., Maitaining a Nursery along with Farming Fetches Better Income, Hindu, 4, 287, pp. 12, December 20, 2014.

Prasad, C. and Byra Reddy, H. N., Extension Strategy for Minimising Risk in Rainfed Agriculture In Rainfed, In Extension Strategy for Minimizing Risk in Rainfed Agriculture edited by Prasad and Das, Indian Society of Extension Education, pp 1-25, 1991.

Prasad, C.," Krishi Vigyan Kendra in Integrated Rural Development", Indian Farming, XXVIII, 7 & 8, pp. 85-89, 1978.

Prasad, R. B., Singh, M. and Singh, S., "Impact of Psychological Factors on Quality of Working Life in Krishi Vigyan Kendra", Maharashtra Journal of Extension Education, XVIII, pp. 262-264, 1999.

Raj, S. Michal., Ram, Nagesh. and Sarang, N., "Success Models of Farm Women in Fish farming", Indian Farming, 48, 11, pp. 23-29, 1999.

Rai, A. K., Thakur, P. and Sharma, R. C., "Role of Front line Demonstration in Transfer of Linseed Technology", Maharashtra Journal of Extension Education, XVIII, pp. 277-278, 1999.

Raj, S. Michal, Ram, Nagesh. and Sarang, N., "Success Models of Farm Women in Fish Farming", Indian Farming, 48, 11, pp. 23-24, 1999.

Raj, S. Michal., Singh, Srojenndra. and Ahmed, S. K. Zaamir., "Farm Women in Cultivation of Vegetable of Crop – A Success Story", Indian Farming, 48, 9, 1998, pp. 31, 1998.

Ram, Ganga. and Mathur, P.N., "Training Priorities of Krishi Vigyan", Journal of Extension Education, XVII, 1 & 2, pp. 48-53, 1981.

Ramasubramanian, M., Senguttuvan, T. and Dhamodaran, "Impact Analysis of Interventions of Krishi Vigyan Kendra on the Composite Crop Culture in Cauvery Delta Region of Tamil" Journal of Extension Education, 23, 1, pp. 4585-4591, 2011.

Ramakrishnan, K. and Rathakrishnan, T., "Impact of KVK Training on Operational Skills Required by the Trainees – An Analysis", Journal of Extension Education, 25 2, pp. 5062-5067, 2013.

Rao, D Uma Maheswara. and Chndrasekhara, P., "Productivity Enhancement in Groundnut through From Line Demonstration in Krishi Vigyan Kendra, Visakhapatnam", Indian Farming, 60, 4, pp.19-21, 2010.

Singh, Madan and Prasad, R.B., "Quality of Working Life in KVKs of Bihar: A Multivariate Analysis", Indian Journal of Extension Education, XXXIII, 3 & 4, pp. 115-120, 1997.

Singh, R.N., Bhagwan, Singh and Saha, D.K., "An Evaluation of Farmers 'Training Programme of Agriculture Technology Management Agency (ATMA) – A Case Study' sent for Publication to Agriculture Extension Review, 2011.

Saini, G.S. and Pandita, B.K., "Attitude of the Training Staff of Krishi Vigyan Kendras Towards its Functioning", Indian Journal of Extension Education, XXXIII, 1 & 2, pp. 68-71, 1997.

Sharma, R.P., "Evaluation of Training Programmes of Krishi Vigyan Kendra", Journal of Extension Education, 13, 2, pp. 3309 -3313, 2002.

Sharma, M.I., Chauhan, M.S. and Sharma, P.N., "Impact of Krishi Vigyan Kendra on Maize Growers", Maharashtra Journal of Extension Education, XVI, pp. 335-336, 1997.

Singh, Baldeo, and Das, P., "Transfer of Technology in NEH Region", Indian Farming, XXXIV, 2, pp. 17-22, 1985.

Saxena, A.K. and Gour, Shashi, "Utility of Krishi Vigyan Kendra's Training Programme", Maharashtra Journal of Extension Education, XVII, pp. 6163, 1999.

Sukhla, A.N., Training Needs and Strategy for Agriculture, In Extension Strategies for Rainfed Agriculture, 1991, Edited by C, Prasad and Das, P. Indian Society of Extension Education, International Fund for Agricultural Development, Indian Council of Agricultural Research, pp. 357-370. 1991.

Sahu, Ram Prakash. Study of KVK of District Unnao (U.P.) with Special Reference to Dissemination of Wheat Technology, Ph. D. Thesis, B. H. U., pp.1-170, 2005.

Singh, Neeraj, Impact of Krishi Vigyan Kendra on Adoption of Modern Practices of Paddy Cultivation in Banswara District, M. Sc. Thesis, Rajasthan Agricultural University, Bikaner, pp. 1-152, 1994.

Singh, Arvind Kumar, Impact of Krishi Vigyan Kendra in Adoption of Mustard Production Technology among the Farmers of Nadbai Panchayat Samiti of Bharatpur District, Rajasthan, M. Sc. Thesis, Rajasthan Agricultural University, Bikaner, pp.1-115, 1996.

Singh, N. P. and Gill, S. S., "Training Needs of Farmers", Indian Journal of Extension Education, XVII, 1 & 2, pp. 66-72, 1982.

Singh, J.N., Tyagi, K.C. and Sohal, T.S., "Training Needs of Dairy Farmers", Indian Journal of Extension Education, XV, 3 & 4, pp.77-82, 1979.

Sunil, N.K., Manjula, N. and Hirenvenkanagoudar, L.V., Role of KVK in Technology Transfer to Create Awareness on the Vermicompost, In National seminar on Appropriate Extension Strategies for Management of Rural Resources, December 18-20, 2007, Dharwad, pp.2007.

Sunil, N.K., Manjula, N. and Hirenvenkanagoudar, L.V., Impact of KVK Training on Adoption of Vermicompost Technology of Participants In North Karnataka, In National Seminar on Appropriate Extension Strategies for Management of Rural Resources, December 18-20, 2007, Dharwad, pp. 94.

Sharma, Preeti. "Extent of Satisfaction of Women Trainees Regarding Training Courses Organised by Krishi Vigyan Kendras of Punjab", Rajasthan Journal of Extension Education, XI, pp.107-109, 2003.

Sharma, Preeti and Mahajan, N., "Status of Women Training Programmes and Infrastructure Facilities in KVKs of Punjab", Indian Journal of Extension Education, XXXX, (3 & 4), pp. 68-74, 2004.

Soni, R.L. and Bhimawat, B.S., "Effectiveness of Institutional Training Programmes for Tribal Farmers", Rajasthan Journal of Extension Education, 10, pp.141-143, 2002.

Singh, Madan and Singh, M.P., "Acquisition of Skill in Rice Farming – A Study in KVK System", Indian Journal of Extension Education, XXXI, 1 to 4, pp.84-86, 1995.

Sripal, K.B. and Kumar, R. Venkatta, "Impact of Krishi Vigyan Kendra Trainings, in National Conference on Extension Education for Sustainable Development, Mumbai, December 29-31, 1998, pp. 26.

Veeraih, R.A., Daivadeenam, P. and Pandey, R.N., "Knowledge and Adoption Level of Farmers Trained in Krishi Vigyan Kendra about Groundnut Cultivation", Indian Journal of Extension Education, XXXVIII, 1 & 2, pp. 58 -63, 1998.

CHAPTER - 9

Conclusion

In the preceding chapters of the book, an attempt has been made to describe four focused attempts designed effectively to bring agriculture technology and knowledge to farmers at their fields and farms viz., (i) National Demonstration Programme (NDP), (ii) Lab to Land Programme (LLP), (iii) Operational Research Project (ORP) and (iv) Programmes carried out by Krishi Vigyan Kendra (KVK). These were devised at different times in response to demands made on research organizations for agriculture development in particular and rural development that follows from it. All these systems have a sound conceptual frame work. This chapter does not propose to present an academic discussion on theoretical differences among these programmes. The following conclusions relate to the various aspects of farmers' education and extension programmes that have been initiated by the Government of India since independence.

NATIONAL DEMONSTRATION

Agriculture Productivity

The strength of farmers' education lies in the availability of relevant and meaningful agriculture technology at research stations. In early 60s Agriculture scientists were for the first time involved in imparting education to farmers about agriculture technology through the National Demonstration Programme (NDP). The programme gave positive results. There was significant increase in yield of various crops in demonstration plots. This shows the availability of scientific know-how of agriculture for the farming community that can bring development of agriculture. The agriculture scientists showed with this programme how it was possible to increase crop yield; as there is a wide difference in crop yield between demonstration plots and farmers' plots. Hence, priority should be given to farmers' education about scientific method of agriculture.

Knowledge and Adoption of Improved Agriculture Practices

Knowledge and adoption of new technology by farmers is a reliable measure of the effectiveness of the extension work. The farmers covered by the NDP possessed better knowledge and more adoption of improved agriculture practices of crops than those of control villages. Adoption of high yielding varieties and use of chemical fertilizer and plant protection chemicals increased considerably among the farmers covered by the NDP as compared to the quantity used by them before the NDP.

Implementation of the Programme

Under the NDP demonstrations concerning improved agriculture practice were conducted at farmers' fields. Such demonstrations were one of the important extension methods for educating the farmers about agriculture technology. The success of such demonstration depends to a major degree on how effectively they are conducted. The studies reported critical observations on the implementation of the NDP. They focused on aspects such as the role performance, perceptions and understanding of agricultural scientists regarding methods of organizing demonstration, extension methods, publicity, and use of mass media and prescribing dosage of inputs.

There are several activities for conducting the demonstrations of scientific agriculture at farmers' fields. These include selection of appropriate technology, farmers and demonstration plots, farmers training on selected agriculture technology, conducting demonstration in fields, publicity of the success of demonstration among farming community. Each activity should be implemented in letter and spirit.

Covering a Variety of Crops and Practices in Field Demonstration

The demonstrations were mostly conducted on improved practices of crop cultivation. However significant time was not given to pulse, fiber and oil seeds. Commercial crops were also not taken on a large scale. Horticulture, fodder and plantation did not find a place to the desired extent. Improved animal husbandry practices were not covered in the NDP.

There should be a close contact between agriculture scientist and farmers for effective extension work. This benefit the farmers as well as the agriculture scientist. The agriculture scientist can have full knowledge of farmer's background and of the resource available to them and local problem. This will enable him to relate his scientific work effectively to the farmers actual needs. This has implication for scientific establishment in terms of modifying their agenda. The agriculture scientist should devote some percentage of his time to farmers' education and their knowledge of scientific agriculture. The agriculture scientists should also be provided facility for extension programme.

Farmers' participation in the programme is crucial for success of the programme. It was reported that farmers' interest was less towards learning technical know-how of agriculture. They were more interested in obtaining free seed and fertilizer. In fact agriculture inputs – seed, fertilizer and pesticide – are essential for agriculture productivity. However, there is a need to change the mindset and action of farming community to get significant production from agriculture inputs. It is possible only to put first priority on farmers' education about scientific agriculture. Research institute should allocate sufficient budget for extension activities for farmers training, group meeting, farmers fair, field day and preparation of teaching aids.

There were also some deficiencies in the management of NDP. There were delays in the supply of seeds and plant protection measures. This is of serious concern. The farmers who do not get timely agriculture inputs may lose their trust in the agriculture scientists and research institutes. As is known, negative news travels faster and wider. Other weaknesses were inadequate and untimely availability of funds for the programme, lack of transport facility, and lack of staff. These have implications again for the research institutes. It is, therefore, necessary for the research institute to take such steps as may be needed to rectify the system for effective implementation of the programme and acquire credibility in the eyes of the farming community.

Evaluation

Evaluation and monitoring is essential to understand how far the programme is functioning. Modifications may be done on the basis of evaluation to meet the changed condition. In addition to self evaluation external evaluation should also be conducted.

OPERATIONAL RESEARCH PROJECT (ORP)

The Operational Research Project was started in 1974 on the recommendation of the working group of the ICAR, New Delhi. It aimed at giving adequate attention to human aspects in all its totality. The ORP is an approach to the use of available resources – soil, water, plants, animals and man – in such an integrated manner that entire rural community including even the children is involved in the process of rural agriculture development. Sharma (1978) studied the impact of the ORP on - (1) Modern technology of milk production and crop production,(2) Reclamation of alkali soil, (3) Composite fish culture and (4) Integrated control of cotton pest undertaken by research institutes. He reported a significant increase in productivity of crop, milk and fish in the areas due to improved agriculture practices. Chauhan *et al.* (1988) made case study to assess the impact of improved agriculture practices carried by the ORP of Central Arid Zone Research Institute, Jodhpur. They reported that agriculture productivity could be increased even in the arid area endowed with poor natural resources by applying improved agriculture practices. They suggested simultaneous attention should be given to improved practices of sand dune stabilization, pasture development, biogas and sprinkler and drip method of irrigation and rodent control. Arya *et al.* (1994) reported that there was significant change in overall economic status of adopted families of ORP villages as compared to the condition to non- adopted villages.

LAB TO LAND PROGRAMME (LLP)

Increased crop productivity

The Lab to Land Programme was started during the Golden Jubilee Year (1979-80) of the Indian Council of Agricultural Research (ICAR), New Delhi. It is an important programme because its aim is to educate the small and marginal farmers and landless labourers about improved agriculture practices who constitute a very large percentage of cultivators. This programme has made a very positive impact among them. The farmers got significantly more yield and income from improved agriculture practices than that of farmers' method of cultivation. There was transformation in the life of the farmers due to the programme. It was important feature that the scientists selected appropriate improved agriculture practice after having knowledge of local farmers' practice of agriculture and farmers' problem in agriculture. This can be replicated for successful farmers' education of scientific method of agriculture.

Knowledge and Adoption of Agriculture Practices among Farmers

Behavioural changes occurred among the farmers as a result of LLP. The level of farmers' knowledge improved method for paddy, sesamum; green gram and apple cultivation were significantly increased in the LLP villages because of extension effort of the agriculture scientists.

The adoption of improved practices of paddy, castor and seasamum cultivation among the beneficiary farmers was significantly higher than among non-beneficiary farmers.

Data collected by researchers from adopted farm families of LLP about adoption of farm practices before and after the initiation of LLP sows that there was significant increase in adoption of improved seed, chemical fertilizer and seed treatment and moderate increases in plant protection and weed control practices among the farm families. The beneficiary farmers of LLP started using the seed rate as per recommendation and applied more quantity of chemical fertilizer for their crops of groundnut, sunflower and paddy. They were also in position to identify pest and diseases which they were not able to do earlier. They used concentrate, mineral mixture and vaccination whereas before the LLP only dry fodder feeding and traditional method of health checking were in practice

Farmers' Perception of Lab to Land Programme

The study disclosed that overwhelming number of the beneficiary farmers of the LLP had factually correct impression of the programme. In contrast majority of non-beneficiary farmers did not have correct image about the programme. It is essential that villagers should spread among other villagers information about the programme. Extension programmes should be organized for this.

Leaflets on the programme should be distributed among the villages and attractive poster on LLP should be displayed at prominent places in the villages. Print and electronic media should be involved in the programme for dissemination of the information about LLP -its objective and activities, its operational method, usefulness etc. This is essential for success of the programme.

Implementation of Lab to Land Programme

For success of extension programme, it is essential that there should be concentrated effort by the extension agency. By merely conducting a number of demonstrations on improved agriculture practices the programme can not be successful. A limited number of improved agricultural practices should be demonstrated in farmers fields and there should be concentrated effort to make them successful. The farmers who cooperate in extension programme, allow other farmers to visit their fields and spread the message of improved agricultural practices in villages should be selected for the demonstration programmes. Scientific methods of soil preparation, selection of good seed, seed treatment, proper spacing and timely weeding should be given priority in the programme. There is no risk in adoption of these practices since no cost is involved and it is certain that benefits will follow. Farmers can not get full benefit from their spending on inputs like chemical fertilizer, pesticide, insecticide and irrigation unless these practices are adopted. The agriculture scientists should make a scheduled visit to the demonstration plots at the time of operation. He should visit the demonstration plot with the farmer.

Agricultural scientists should conduct the field day with sufficient publicity at demonstration plots to spread the message of improved agriculture practices which are being implemented in the demonstration plot. The demonstration farmer could speak about the improved agriculture practices done in the field to the audience of other farmers after which it should be the agriculture scientists turn to speak. The farmers should be encouraged to raise questions. Leaflet should be distributed for greater impact of the field day. Visual aids – poster and chart should be displayed during the field day. The agriculture scientists may make a report on the demonstration and its impact. The personnel of radio, television and newspaper may be invited to attend the field day so as to give wider publicity to disseminate the message of new agriculture practices among the farmers.

There should be a close linkage between research institutes and mass media – radio, television and print media. The agriculture scientists should use the media to spread information about the LLP and improved agriculture practices for dissemination among the villagers. The name of improved seeds of crops, insecticide and pesticide are in English which the farmers can not pronounce and remember. It should be repeated during entertainment programme in radio and television. The unit of measurement should be the local unit. The progressive farmer should be invited by radio and television to present his views about improved agriculture done in the demonstration plots and how he benefitted from the adoption of improved agricultural practices. There should be adequate coverage of agriculture practices, availability of agriculture inputs and their cost in newspaper before the onset of the monsoon. The information on field problems of agriculture like spread of insects and diseases and their control measures should be covered in newspaper. The personnel of mass media and newspaper should go and stay a few days in the village to know how the farmers carry out cultivate their land, what problems they face in doing so and what functional and situational problems are there in agriculture. This will help them in designing relevant programmes for the farmers.

Organisation of effective training is an integral component of an extension programme. There has been a dearth of studies on the training programmes.

Subsidy on Agriculture Inputs

The amount of subsidy for critical inputs is not sufficient to meet the required input expenditure for the recommended technology. Further, as soon as the LLP programme was withdrawn after two years of operation farmers were reluctant to use chemical fertilizer due to its high cost. It was found that the farmers felt the cost of improved seed was high and also there was no availability of improved seed at the right time. It is therefore, essential that the farmers should be educated about advantages of growing of improved seed. If the cost of transport is reduced the seed will be cheaper. There will be availability of good quality seed in time which is very important for timely sowing of crops. A few farmers should be selected and should be given training for about growing good and improved quality seed in their fields. The ICAR research institutes and Agriculture

University should also produce improved seed for sale to the farmers.

The study also reported high cost of chemical fertilizer owing to which, the farmers were reluctant to use chemical fertilizer when the LLP programme was over. This problem can be minimized if the farmers are educated about composting of cowdung. In India, most of the farmers keep cattle. Thus cowdung is available which can be used for composting. However, it is not so beneficial for crops when it is not well decomposed. Cowdung is usually kept open in a heap. When it is applied in the field, it also becomes a source of termites. The farmers should be educated about composting the cowdung, left over fodder and kitchen material. They can thus reduce quantity of the chemical fertilizer in their fields. Besides maintaining fertility of soil, compost maintains soil moisture for long period which is vital for crop productivity.

The farmers should be educated about soil testing. Agriculture research institutes should make available to the farmers the results of soil tests. More laboratories should be established so that the farmers should avail of the facility.

Farmers buy their requirements of agriculture inputs from private dealers. Private dealers are the source of seeds; fertilizer and pesticide for farmers. They do not have adequate knowledge of agriculture. Short trainings should be organized for them by research institutes of ICAR and Agriculture Universities. There is a suggestion that licenses should be give to unemployed agriculture graduates who can contribute in educating the farmers about new agriculture practices.

Activities of the Programme

Farm holdings are decreasing day by day due to increase in population because of which the income of the farmers is also decreasing. Several farmers take recourse to borrowing and get trapped in debt. Under these circumstances the best possible solution would be to adopt integrated farming system where along with field crops, animal husbandry and horticulture crops including fruit, vegetable and spices component become important source of income for the farmers. The choice of crop, livestock and horticulture crop will depend on factors like irrigation, rainfall, size of holding and available resources of the farmer particularly working hands when needed. The guidelines for LLP emphasize the need for extension activities in such areas as animal husbandry, sericulture and apiculture since these activities provide supplementary income to the small and marginal farmers. These activities also minimize the effect of drought. Though major emphasis in the LLP was on crop cultivation, there is nevertheless need to adopt the integrated farming system

At present the research which is being done is mainly crop specific, livestock specific and horticulture specific research is a separate component. The recommendations of these separate research findings can not solve the problem of farmers who need to adopt integrated farming system approach. Therefore, there should be research on integrated farming system and suitable integrated farming system models may be developed for different agro-situations.

Short Duration of the Programme

The LLP is implemented in villages for two years. After two years other villages are chosen for the programme. It is a very short-duration for the programme. Within this short period the agriculture scientists fail to acquire in-depth knowledge of the farming community there and to local farm practices. Initially the farmers do not cooperate in the programme. When the farmers see the result of improved agriculture practices carried out at the fields they develop confidence in the programme. There are several agriculture practices for example agroforestry which show their impact after considerable period. The programme should have five years period for optimum results. Such a span of years will enable the agricultural scientists to know the farmers background, resources available to them and local agriculture problems and will enable them to relate improved agriculture practices effectively to their lives. Such village will in time become a model for the area for its modern agriculture practices. The neighbouring villages will learn about the agriculture practices and be motivated to adopt them.

KRISHI VIGYAN KENDRA (KVK)

Following the recommendations of the education commission, the Government of India launched KVK in the country in 1974. The KVKs provide the government presence in the country for the farmers' education about scientific agriculture developed at research institutes for agriculture development.

Training

Under the KVK, there is an organized effort for farmers' training in improved agriculture technology. At national level, between 2007-08 and 2013-14, the KVKs organized 2,59,645 courses for 77.04 lakh farmers out of which 17.78 lakh were women (24 per cent). The courses covered a wide range of subjects. The largest percentage was of courses on crop production, horticulture, plant protection, livestock and empowerment of rural women.

Effectiveness of training programme is reflected in the cognitive and affective reactions of the farming community. The pre and post assessment methods of training evaluation revealed a tremendous increase in the level of knowledge of improved agriculture practices among the farmers.

It is reported by the study that the farmers have a high level of training need. The farmers took a great interest in difficult subjects *viz.* chemical fertilizer, plant protection, farm machinery and veterinary service.

KVKs training programme benefited the farming community by upgrading their knowledge of agriculture technology. There are few studies on skill development of farmers about agriculture technology. If the farmers can not use the knowledge due to lack of skill, they may lose trust on the training programme and organization. Merely having more number of courses and trainees will not do. The quality of training programme should be increased substantially. In fact, the imparting knowledge and skill to the farmers is not an easy task since learning by the farmers is voluntary. As the trainer has a key role in the training programme, he should have a genuine desire for village welfare and in depth of knowledge of improved agriculture technology. The number of training course and topic should be based on the farmers' needs. For this, it is essential for the trainer to have full knowledge of farmers' background and of resources available to them and of local problems faced by the farmers in cultivation. The trainer needs to understand local agriculture practices by discussing with farmers and observing things in the fields. This will enable him to relate the improved agricultural practices effectively to their lives.

The trainee's size should be of ten to twenty persons to successfully impart knowledge to the farmers. A limited number of agriculture practices should be undertaken in training session, but they should be well covered. The farmers learn better, faster and easier through seeing and doing. For example, the trainer could show the samples of seeds in the training session. It would be even better to show them in the field and explain their characteristics and cultivation practices. The trainer should ask the farmers to restate the content of the subjects taken up. If the farmer feels difficulty in reproducing the subjects, it should be repeated so that the farmer can fully understand the content.

Practical demonstrations provide skill (competency) in the trainees of new agriculture practice. There should be limited number of trainees, even a single farmer for practical on agriculture. The practical should be properly planned and organized. The equipments required for conducting practical should be available and checked well in time.

The farmers should be provided pamphlets on the subjects dealt in the training session. The specimen or other visual aids should be provided. The trainer should consider time, duration and place of the training while planning the training programme.

The trainer should evaluate the training programme from time to time with respect to knowledge and skill gained by the farmers.

Knowledge of Improved Agriculture Practices

The KVKs adopt villages and use several extension methods for educating the farmers about improved agricultural technology. At national level 3,54,412 extension activities were organized in the villages adopted by the KVKs benefiting 674.71 lakhs farmers and extension workers between 2007-08 and 2013-14. 6,54,412 activities were organized between 2007-08 – 2013-14 through electronic and print media for wider coverage. The objective was dissemination of scientific know-how and scientific methods of agriculture among the farmers.

The impact of the KVK with respect to gain of knowledge regarding improved agriculture practices for different crops was assessed by researchers who questioned both the beneficiary farmers of KVK villages and farmers of villages not adopted by KVK. It was found that the beneficiary farmers had much higher knowledge than the non-beneficiary farmers regarding improved agriculture practices for crops.

It was found that the farmers did not have adequate knowledge of non-monetary agricultural practices. It is essential that the farmers should have adequate knowledge of these practices for a higher crop yield. For example a proper distance between plants and rows for sowing has been recommended by the agriculture scientists. If in some places the plant to plant and row to row spacing is less and other places it is more, it affects the utilization of soil moisture and plant nutrients by crops. It will result in crops poor growth and poor yield. Similarly improved method of weed management has been recommended. If weed is not well managed, there will be competition between the crop and weed for soil moisture and chemical fertilizer applied in the field. It will affect crop resulting in a low yield. The farmer should have knowledge of the crucial stages when irrigation is required in different crops. If the irrigation is not available at the critical stage, growth of the crop will be adversely affected. Also crops are attacked by pests and diseases resulting a substantial loss to the farmers. The knowledge scientific methods of storage of farm's produce are essential to protect it from insects and pests.

The beneficiary farmers have a crucial role in educating the farmers about improved agriculture practices since they have more contact with the KVK staff. There is a lack of studies how the beneficiary farmers were selected and how they communicate learned agriculture from KVK in villages. The selection of the beneficiary farmers is an important factor for their effective functioning. The beneficiary farmers who are progressive, cooperate and extend the knowledge learnt from the KVK staff to the farmers in the villages should be selected. A vigorous effort should be made by the KVK staff to help the farmers learn about the latest advances in agriculture technology. Adequate knowledge will motivate and inspire them to adopt the practice which is feasible in their situations.

A large proportion of the farmers are unable to read and write and even literate farmers have low level of education. Therefore message of scientific know-how of agriculture should be told to the farmers in a way that they can understand, accept and act on them.

The scientific know-how of agriculture should be communicated in the local dialect. The technical words viz, name of high yielding variety of crops, chemical fertilizer, insecticide and pesticide are in English. Hence poster should be made on them and should be displayed in common place in villages. Wall poster can also be made on them in villages. The farmers will see them and remember them. Leaflet and pamphlets should be made on scientific know-how of agriculture and these should be provided to the educated farmers. The author of the book had an encounter with young and educated farmers. This has a deep impression in the mind. The farmer had kept newspaper in which agriculture information was discussed. He showed the author and told that on this information he enquired about availability of improved seed of peal millet – CZP-9802 from Central Arid Zone Research Institute, Jodhpur. The educated farmers can spread the information of agriculture learnt from leaflet among the farmers. The dose of chemical fertilizer, insecticide and pesticide should be told to the farmers in local unit instead of metric unit.

There are several programmes for the farmers in the country which are implemented by extension agencies of agriculture and animal husbandry. There should be a linkage between extension agencies and KVKs. Similarly there should be linkage between KVKs and non-government agencies. The KVKs should provide training to the personnel of agriculture input agencies.

Adoption of Improved Agriculture Practices

Adoption of improved agriculture practices of cultivation of wheat, paddy, mustard pear millet and groundnut was studied by different scholars. The adoption of improved agriculture practices was not same among the crops. The mean percentage adoption score of the crops among the beneficiary farmers shows that only one fourth of improved agriculture practices of groundnut was realized, half of the improved agriculture practice of paddy was realized and better levels in case of pearl millet, wheat, mustard and henna.

In case of the non-beneficiary farmers, the adoption was very low as compared to the beneficiary farmers. The realization of improved agriculture practices was only one fourth in groundnut, one third in case of paddy, mustard and henna and half in wheat crop.

Adoption of improved agriculture practice of crops is a very dynamic process. Knowledge among the farmers about improved agriculture practices is essential. However, timely supply of quality inputs (seed, fertilizer and plant protection chemical) in adequate amount is essential to fulfill the farmer's demands. It is also essential that there should be remunerative price for farm produce. By getting better price, they would invest more in agriculture, resulting in more agriculture production. At the same time it will also enhance the socio – economic condition of the farmers which constitute major part of the population of the country.

Demonstration Farm

The KVK must demonstrate scientific method of crop rearing, animal husbandry, horticulture at its campus to convince the farmers. The KVKs should produce seed of crops and grass, and saplings for the farmers. This will help generate income of the KVKs.

Area of operation of the KVK

There is one KVK in each district in the country. This is much too large an area for effective operation by one KVK. Hence there should be two KVKs in each district in the country

SUGGESTION FOR FUTURE RESEARCH

Agricultural scientists of ICAR, New Delhi and those at agriculture universities have under NDP, LLP, ORP and KVK programmes and interacted with farmers in the villages and formally trained them in the scientific techniques of agriculture. They also conducted studies to assess the impact of the programmes. This work is based on their observations regarding the success of their efforts to educate/train farmers on improved agriculture practices, their discussions with the farmers and collection of data from the farmers through survey and case studies. The work done in the field has provided valuable experience which can be used for strengthening the programmes for farmers' education and also provide guidance to policy makers and extension workers in the country and abroad. There has been no independent research on these programmes. More research should be done on the programme by independent agencies also. The studies also need to have larger numbers and wider geographical coverage of respondents.

Majority of the studies were on crop productivity and knowledge and adoption of improved agricultural practices. There was dearth of study on skill development of farmers required for adoption of improved agriculture practices. Further study should be undertaken on skill development among the farmers. Besides this, the study should be conducted on functioning of the beneficiaries of the programme.

There should be more study on implementation of the programmes and the personnel involved in the programme.

■■■■